Praise for

THE QUANTUM & THE DREAM

"Doug Grunther is a radio bodhisattva"

—**ROBERT THURMAN,** Noted American Buddhist writer
and first Westerner ordained by the Dalai Lama

"Written in a clear, engaging, and personal style, and drawing upon numerous thought-provoking sources in science, psychology, and philosophy, Doug Grunther's *The Quantum and The Dream* addresses a number of important topics of great relevance to our contemporary times including right brain intuition, the emergence of a global network of communication and cognition, and our human relationship with the complex coordinative intelligence of Nature and the Earth."

—**THOMAS LOMBARDO,** Director, The Center for Future Consciousness

"In Doug Grunther's sweeping, deep and delightful new book, *The Quantum and The Dream,* we are led on a journey through much of the 20th century's innovative and challenging new ideas. From quantum mechanics and psychoanalysis, to the Internet, and an emerging holistic view of nature and consciousness, Grunther provides an integrative and positive understanding of our current direction and ultimate integration of seeming opposites. *The Quantum and the Dream* is written in an amusing, clear, easy to read, and ultimately existentially deeply helpful voice that provides an optimistic view of our rapidly emerging future. A must read for all those—and that is most of us—interested in understanding and grounding of our past in our future. Bravo, Grunther!"

—**NEAL GOLDSMITH, PH.D.,** psychotherapist, public speaker, and AI analyst

THE
QUANTUM
&
THE
DREAM

VISIONARY CONSCIOUSNESS, AI,
AND THE NEW RENAISSANCE

Douglas Grunther

Epigraph Books
Rhinebeck, New York

Paperback ISBN 978-1-960090-56-0
eBook ISBN 978-1-960090-57-7

Book design by Colin Rolfe

Epigraph Books
22 East Market Street, Suite 304
Rhinebeck, New York 12572
(845) 876-4861
epigraphps.com

CONTENTS

SECOND SHIFT

From Printed Page to Digital Screen

THIRD SHIFT

From the Age of Information to the Age of Recognition

INTRODUCTION

A New Renaissance?

How can that be, with the constant bombardment of bad news, the world burning and flooding from escalating climate change, violence and authoritarianism on the rise, and AI seen as a malignant force which will eventually supplant humanity?

But underneath all the fear and anxiety is a deep pattern of transcendent change which began over a century ago, and which continues to emerge from the cataclysmic turbulence sweeping the planet here in the twenty-first century.

This deep pattern is what we'll explore in the journey ahead.

First, a word about **Synchronicity.**

The term "synchronicity" was created by Carl Jung, who described it as two or more events which clearly didn't cause one another, but are too meaningfully connected to be a mere coincidence. His more poetic description of synchronicity is *"a falling together in time."*

As we will see, Jung collaborated with one of the most brilliant quantum physicists in the world. Together, they were convinced that synchronicities occur not just at the individual level, but at the deeper and more expansive level of a collective unconscious, one shared by all of humanity. We'll see how this synchronicity generated two parallel, fascinating descents: one into the dark, creative center of the unconscious mind; the other into the dark, creative center of the sub-atomic quantum realm.

And so, I invite you on a journey, one which starts with a paradigm-shifting synchronicity that emerged in the first year of the twentieth century, just as the modern age was unfolding.

A WORD ABOUT YOUR GUIDE ON THE JOURNEY

I have been a seeker all my life, a passionate researcher, and a synthesizer of ideas, curious about the depths of the human mind. **The Quantum & the Dream** is the brainchild of my forty-plus years as a radio talk show host in extended conversation with some of the most innovative thinkers on the planet—psychologists, philosophers, spiritual teachers, scientists, and cultural historians. These people opened my eyes and my mind, and led to my recognition of a never-before-acknowledged pattern pointing us towards this New Renaissance slowly emerging.

We humans are wired to recognize deep patterns. **The Quantum & the Dream** synchronicity demonstrates how three forms of intelligence—Human Intelligence, Computer Intelligence, and the most influential of the three, Nature's Intelligence—are creating evolutionary change which points, not to end-time apocalypse, but to a New Renaissance.

ORIENTATION *for* THE JOURNEY AHEAD

The key to evolution is novelty and creative
change. Without variation and novelty, selection
has nothing to act upon.
—JAMES SHAPIRO, BIOLOGIST

We've left the world of logical,
one-step-at-a-time
sequential thought . . .

We have entered the age of the
Synaptic Jump Cut
& Quantum Leap

Billions of ideas, intuitions, products, preferences, insights, and nervous systems are coursing around the planet at electric speed through fiber optic cables, under oceans, over continents, and streaming down from satellites orbiting the earth, weaving a global, hyperlinked matrix connected to over five billion of us and counting . . .

Through digital screens we now have immediate access to the world's greatest knowledge and wisdom.

Advances in medicine and science propel forward at ever faster rates. Deep research that just decades ago would take months and years can now be done within hours and days.

Artists, professionals, and innovative thinkers can exchange information and start to pollinate new perceptions instantly around the globe.

At the same time, fear and anxiety from the incessant waves of the twenty-four-hour news cycle, fake news, trolling, social media obsessions, and the constant flood of aggressive ads compelling us to consume, all grab for our attention with disturbing levels of cognitive dissonance.

As Kevin Kelly, cofounder of *Wired* magazine, points out, "We're morphing so fast that our ability to invent new things outpaces the rate we can civilize them."

To get a sense of balance, we will be looking at three key shifts in consciousness—one already well under way, the other two starting to appear—with the potential to bring about a rebirth of vision and values moving forward.

These three key shifts are:

I From the Left Hemisphere of the Brain to the Right Hemisphere
II From Printed Page to Digital Screen
III From the Age of Information to the Age of Recognition

The first shift in consciousness, to the right hemisphere of our brains, is where imaginative, boundary-expanding messages from the creative unconscious first percolate up into conscious awareness, increasing the chance for new insights to occur. Modern advances in neurology, which can see directly into the brain in lab testing, confirm this.

When we dive into the second shift, we'll see how the underlying process taking place between our brains and the illuminated, hypertext-connected images on the digital screen, while disorienting at first, closely reflects the way our brains actually generate new breakthrough perceptions.

The third shift, from the Age of Information to the Age of Recognition, looks at how the superior ability of computers to research, organize, and analyze information frees us up to explore our deep connection to Nature's Intelligence, giving us a bigger picture of who we are, how we got here, and where we are potentially heading.

You are about to visit a number of insightful visionaries as they explore the creative depths of the human mind and Nature's incessant beat of cycles in sync.

Welcome to **The Quantum & the Dream.**

FIRST SHIFT
FROM LEFT BRAIN HEMISPHERE TO RIGHT

Imagination is more important than knowledge. For knowledge is limited to all we know and understand, while imagination embraces the entire world, and all there ever will be to know and understand.
—ALBERT EINSTEIN

No paradox, no progress.
—NIELS BOHR

DOORWAY *into* THE MODERN MIND

*In the 20th century the modern materialistic
worldview began to unravel in the face of
scientific and psychological developments. It led
a number of thinkers to consider that the human
psyche may be more involved, in some mysterious
way, with the observed properties of matter.*
—PHILOSOPHER/SCIENTIST THOMAS MCFARLANE

As the nineteenth century was coming to an end science had made leaps and bounds in discovering how the universe worked. Through the genius of Isaac Newton, who discovered the mechanized laws of cause and effect, and the influential philosophy of Rene Descartes, who described whatever could not be seen by the rational mind as dangerous and misleading, a convincing set of fixed principles was established which described the world we lived in as predictable, logical, and ruled by clear and understandable physical laws.

The scientific and philosophic establishments were feeling immense confidence. They had created the awesome technologies of the Industrial Revolution, and a nearly complete description of nature was at hand.

Or was it?

In 1900, the first year of the twentieth century and the Modern Era, the idea of a mechanical, materialist, predictable universe was about to come apart at the seams, shaken to its core. What followed was an enormous paradigm-shift in perception as to who we are and how we connect to the external world.

Throughout the journey ahead, as we encounter this new reality, this slow, but steady "unraveling" of the materialist worldview, there's a significant benefit to understanding how modern neurological evidence points to the distinction between the left and right hemispheres which comprise the frontal cortex of our brains.

A TALE OF TWO HEMISPHERES

Anatomically, the left and right hemispheres of our brains are divided by a band of neural tissue called the corpus callosum. These two hemispheres are primarily what distinguishes us as the most intelligent species on the planet and the most destructive. Referring to the evolution of these most important frontal lobes, Dr. Iain McGilchrist, who has an extensive background in neurology, psychiatry and philosophy, posited that if the evolution of our frontal lobes made us the most destructive of animals, it is also responsible for turning us into the "social" animal, one with an extensive collaborative dimension beyond one's immediate family.[1]

The world view that began to unravel at the beginning of the twentieth century was one dominated by the left hemisphere of the human brain, while the emerging new paradigm revealed evidence that, as McFarlane said, "the human psyche may be more involved, in some mysterious way, with the observed properties of matter," and can only be discerned with any depth with a shift to a more right-hemisphere perception.

The reason is not, as many sources would have us believe, that the left hemisphere of our brains isn't creative. It can be quite creative. But the left hemisphere of our brains can only be creative within a fixed boundary of certain, step-by-step assumptions. To get a bigger picture of the enormous changes being felt in this new, digital, globally-connected age—a seismic change initiated under the surface of public knowledge in the year 1900 and extended outward with greater force ever since—requires that we think beyond the known boundaries of the entrenched materialist/industrial-age paradigm still dominant today. We're about to see that the right hemisphere of our brains is wired for this purpose.

A QUICK THOUGHT EXPERIMENT

Here's a short, easy thought experiment to experience this important shift in consciousness:

1. Eyes closed, in your mind's eye visualize you're standing about twenty yards from an oak tree in a forest.

   With your mind's eye as a camera, zoom in and get a close-up of the tree so you can identify specific details such as the size of the trunk, the number of branches, the color of the leaves if there are any. Get a good picture of the individual parts making up the tree.

   Open your eyes.

2. Eyes closed, again, standing about twenty yards from the tree, this time take in the whole image of the tree. Let the whole image imprint on your mind.

   Now, like a drone camera, elevate your perspective until you can see above the tree and beyond to how your individual oak tree is positioned among the forest of trees. Take in the total view. Then lower the drone camera in your mind's eye back to the ground again.

You just experienced the shift from the left hemisphere of your brain (wired to understand objects by breaking them down into smaller parts) to the right hemisphere (wired to understand an object by seeing it as a "whole").

When you were close-up to your imagined oak tree, taking note of its details, you were using the special wiring of the left hemisphere of our brain which is excellent for collecting data and focusing intently on details. But you literally couldn't "see the forest for the trees." Elevating your perspective with the imagined inner-eye drone camera requires a shift to the right hemisphere, wired to give a broader context, a "bigger picture" for understanding the complex of relationships which provide deeper meaning.

A recent *Scientific American* article confirms that a deeper creativity resides in the right hemisphere, since its specialized characteristics make it, "the seat of curiosity, synergy, experimentation, metaphoric thinking, playfulness, solution finding, artistry, flexibility, synthesizing and in general, risk taking. In addition, it is likely to be opportunistic, future oriented, welcoming of change, and to function as the center of our visualization capability."[2]

There's one more quick thought experiment involving your imagined oak tree which sets the stage for the fascinating journey ahead.

1. In your mind's eye, visualize you're standing about twenty yards from your tree. With your mind's eye as a camera, imagine you can lower your vision beneath the ground of your tree to watch its underground roots absorbing nutrients from the dark soil and percolating them up into the trunk, then further up to any branches and leaves. Return your perspective to ground level.

This shift in perspective to underground is analogous to consciously accessing the deep, enriching psychological "nutrients" inherent in the unconscious, a shift requiring the boundary-expanding capabilities of the right hemisphere. As pointed out by the extensive research and insights provided by Dr. Iain McGilchrist, "The major difference between the hemispheres lies in their relationship with the unconscious mind, whether that means the dream state . . . or what we experience or bear in mind without being aware of it. Whatever does not lie in the center of the attentional field, where we are focused, is better yielded by the right hemisphere, and the left hemisphere can sometimes show surprising ignorance of it."[3]

With this shift from left hemisphere to right in mind, especially its potential for exploring the fascinating, deeply creative threshold space between the conscious and the unconscious, let's travel into **the Quantum & the Dream** synchronicity as the twentieth century was about to unfold.

FREUD INTERPRETS DREAMS &
MAX PLANCK DISCOVERS *the* QUANTUM

Max Planck was a pragmatic German scientist who had been commissioned by several electric companies to create a more efficient light bulb. But scientific discoveries have a tendency to generate unintended and totally unexpected consequences. Such a consequence greeted Planck.

He found it enormously disconcerting.

Science had long accepted the nineteenth-century discovery by James Clerk Maxwell that light consists of a steady stream of continuous waves: it was a cornerstone of science. But Planck's efforts to create a more efficient light bulb seemed to indicate that under certain conditions light was generated and received in the form of individual particles, not waves. He described these particles as "discrete bundles of excited energy."

But how could that be? Particles ("discrete bundles") are individual entities with clear, discernible boundaries, like billiard balls. (Planck called these individual units "quanta, the Latin term for "how much.") Waves on the other hand are underlying forces of energy which spread outward and have no clearly definable boundaries.

Planck tried to grasp the meaning of the apparent paradox he had uncovered: Light could exist as waves in some instances, discrete particles in others. He had found a way to solve a practical problem, creating a more efficient light bulb, but the apparent wave-particle paradox disrupted his need for clarity and consistency.

Max Planck was at that time a very left-hemisphere thinker: sharply focused, analytical, searching for clear, limited boundaries.

Danish science historian Hele Kragh describes Planck as the archetype of the classical mind who, throughout his career as a physicist and statesman of science, maintained that the goal of science was a world picture built on absolute and universal scientific laws. Planck believed such laws not only existed but reflected the inner workings of nature, an objective reality "where human thoughts and passions had no place."

Freud's publication of *The Interpretation of Dreams*, also in 1900, set off a wave of recognition that reverberated in medicine, psychology, philosophy, music, literature, and the arts. It was the seed of an exciting new way to explore the depths of the human mind: psychoanalysis.

In contemporary times, psychoanalysis has morphed into the less formalized (and more widely practiced) depth psychology, a term I prefer as it helps erase the misperception that delving into the creative depth of the unconscious requires formal and long-term psychoanalysis. Depth psychology is very democratic. Anyone open to their dreams, reveries, intuitions, thought experiments, and deep contemplation can access and reap enormous benefits from the creative unconscious.

As Max Planck discovered those "excited bundles of energy" he called "quanta," Freud, that same year, 1900, gave new meaning and importance to the "excited bundles of energy" inherent in the dreams percolating up from the unconscious mind of sleep into the consciousness of waking awareness.

As we shall see, Freud's emphasis on uncovering the powerful forces of the unconscious mind, a descent into the depths of our human psyche, mirrors the descent into the dark mysteries of the sub-atomic world initiated by Planck's discovery of the wave-particle paradox.

It's important to note, however, that Planck's extremely rational left-hemisphere thinking was in tune with virtually the entire scientific mindset at the time. The international scientific community heralded the practical benefits of Planck's discovery at the same time it "repressed" any urge to contemplate the deeper meaning of the apparent wave-particle paradox.

But the repression of deeper meaning can only hold up to a point—the unconscious mind eventually has its say.

Five years after **the Quantum & the Dream** synchronicity emerged in the year 1900, an imaginative, right-hemisphere-oriented thinker, prone to creating playful thought experiments and fascinated by one of his childhood dreams, would make two discoveries that would rock the core of science, philosophy, psychology, literature, art, music, and popular culture.

EINSTEIN CHASES *a* LIGHT BEAM *and* REFLECTS *on a* WILD DREAM

My entire career has been a meditation on a
dream I had when I was eleven years old.
—ALBERT EINSTEIN

In 1905, Einstein, inspired by Planck's uncovering the wave-particle paradox of light, is deep in thought. Unable to get a position in academia, he is working in a Swiss patent office which gives him plenty of time to continue what he enjoyed so much from the time he was a young teenager—creating imaginative thought experiments. So, he is not working in a "material" laboratory of chemicals, test tubes, and Bunsen burners. His experiments played out, not in a lab, but in the right hemisphere of his brain.

Einstein had a playful love of dreams, reveries, and imaginative leaps into the unknown. One could compare his creative thinking process to the attraction of those amusement-park rides where we feel both the thrill and disorientation of being hurled off balance. As Einstein said of Planck's discovery of the quantum paradox, it "was as if the ground had been pulled out from one, with no firm foundation to be seen anywhere, upon which one could have built."[4]

Biographer and journalist Walter Isaacson describes the thought experiment Einstein credits as a key to his discoveries:

"Einstein tried to picture what it would be like to travel so fast that you caught up with a light beam. If he rode alongside it, he later wrote, 'I should observe such a beam of light as an electromagnetic field at rest.' In other words, the wave would seem stationary. But this was not possible according to Maxwell's equations, which describe the motion and oscillation of electromagnetic fields."[5]

The conflict between his thought experiment and Maxwell's equations caused Einstein "psychic tension," he later recalled, and he wandered around "nervously, his palms sweating." Such psychic unease, the kind that leads to an imaginative leap of insight, is much more conducive to the expansive capabilities of the right hemisphere of our brain than to the explicit-seeking, concrete result searching proclivities of the left-hemisphere.

Here is Einstein's description of the dream he had at eleven:

> *I was sledding with my friends at night. I started to slide down the hill but my sled started going faster and faster. I was going so fast that I realized I was approaching the speed of light. I looked up at that point and I saw the stars. They were being refracted into colors I had never seen before. I was filled with a sense of awe. I understood in some way that I was looking at the most important meaning in my life.*[6]

PLAYING WITH A PARADOX

A second Einstein paper written in 1905, inspired by his fascination with Planck's discovery of the quantum, was the first ever submitted on a quantum theory of light (it was for this paper he received a Nobel Prize, not the paper submitted the same year revealing the special theory of relativity).

Einstein's curiosity led him deeper into the mystery a few years later.

In 1909, while considering light's momentum, he became more

and more curious about the implied wave-particle paradox that Max Planck and the scientific community continued to ignore. This "duality" of light's nature had never been fully described before.

Einstein's special theory of relativity is based on light consisting of waves. His paper on the quantum theory of light is based on light consisting of particles. Planck's unintended discovery in 1900 implied this paradox. Now, Einstein was proving it. But how could light be both a particle in some instances and a wave in others?

Instead of trying to resolve the paradox, Einstein embraced it, allowing his mind to sink into it and notice what emerges from the deeper levels of the unconscious mind. He worked with light as a wave, and then again as a particle, picking the attribute he needs to resolve each problem in turn.

This is a great example of the shift from primarily left-hemisphere thinking (tightly focused, understanding by breaking things down into smaller parts to figure out how things are put together) to right-hemisphere thinking (open to anything which might appear in the imagination, ready to take intuitive leaps into the unknown and see things as a whole).

And Einstein's playful use of dreams, thought experiments, and intuitive leaps speaks to the inner laboratory of the human mind, where the unseen and unknown have the potential to combust into storylines with the power to reveal Nature's secrets (and, as we shall see, offer transcendent visions into the depths of the human psyche).

WHAT HAPPENS TO PAST, PRESENT, AND FUTURE?

We now start to get a feel for what appears at first glance to be Einstein's totally irrational, hard-to-believe statement: "The distinction between past, present and future is only a stubbornly persistent illusion."[7]

This insight gets us closer to the nature of a synchronicity, the "falling together in time" of two or more events clearly not causing one another, yet too meaningful to be dismissed as mere coincidence.

For where in the human unconscious is the line dividing past, present, and future? Einstein, inspired by his childhood sledding dream, his right-hemisphere thought experiment chasing a light beam, and Planck's unintentional discovery of the quantum, weaves them all together into a totally novel mind-tapestry, a leap of perception that would have a profound effect on the future of science, philosophy, psychology, art, and culture.

Here, from his book *The Master and his Emissary: The Divided Brain and the Making of the Western World*, Dr. Iain McGilchrist explains why the brain's right-hemisphere capacity to stretch beyond known boundaries is so essential: "In almost every case what is new must first be present in the right hemisphere, before it can come into focus for the left . . . only the right hemisphere can direct attention to what comes to us from the edges of our awareness. . . . Novel experience induces changes in the right hippocampus, but not the left."[8]

Einstein's discovery of the quantum nature of light will go on to inspire a brilliant group of four future Nobel Prize winning physicists with a deep commitment to ancient spiritual insights to create the most successful theory in the history of science—a theory which will produce the amazing technologies of the twenty-first century while also providing evidence that, as we've previously seen in the quote from science historian Thomas McFarlane "... the human psyche may be more involved, in some mysterious way, with the observed properties of matter."

SHINING *a* LIGHT *on* *the* UNCONSCIOUS

The poets and philosophers before me discovered
the unconscious; what I discovered was the
scientific method by which the unconscious can
be studied.
—SIGMUND FREUD

There is a connection between Max Planck's discovery of "excited bundles of energy" he called "quanta," and Freud, in the same time period, giving new meaning and importance to the "excited bundles of energy" in the dreams percolating up from the unconscious mind of sleep into the consciousness of waking awareness.

But as he himself noted, Freud didn't discover the unconscious.

Henry Ellenberger, in his book *The Discovery of the Unconscious,* points out that the romantic poets in the first half of the nineteenth century anticipated much of what Freud would eventually document.

One of the nineteenth century's most influential philosophers, Arthur Schopenhauer, argued that most of our thoughts and feelings are unknown to us but that the reason for this is an unconscious repression.[9]

And the great American psychologist William James, among others, spoke and wrote extensively about the unconscious system in mind and body in the 1890s.

But it was Sigmund Freud, gifted neurologist and medical doctor, whose book published at the turn of the twentieth century provided scientific methods, including the creative process of "free association" for uncovering previously hidden messages, noting the fascinating and therapeutic connection between the unconscious psyche and enduring mythological stories, those dark, cathartic dramas from ancient Greece such as Oedipus and Electra.

With the publication of *The Interpretation of Dreams,* the unconscious, long repressed and feared by science and psychology, would finally take center stage.

A fair question to ask is why, if Freud is so important to twentieth-century thought and insight, have his methods and theories lost so much of their luster in the new millennium?

While there's considerable reason to criticize Freud's personality (he was autocratic, unyielding, intolerant of criticism, and mired in the sexist Victorian view of women) and to note the discarding of many of his specific methods, what's important for our understanding of **the Quantum & the Dream** is that Freud held up a huge mirror, revealing the extensive reservoir of hidden drives determining our perceptions, thoughts, and beliefs about who we are, and how we connect to each other and to the world at large.

The publication of *The Interpretation of Dreams* just as the Victorian era was turning into the modern era would eventually, as we will see, generate an enormous shift in awareness which greatly influenced virtually every aspect of culture.

As artist and art critic Rosie Lesso writes, "Sigmund Freud forever altered the way we see ourselves. Believing our adult behavior is driven by repressed childhood experiences of love, loss, sexuality and death [still resonates] within the arts . . . the creativity and inventiveness of Sigmund Freud's theories continue to fascinate and inspire countless creative thinkers."[10]

JUNG'S BIG DREAM

To me dreams are a part of nature, which harbors
no intention to deceive,
but expresses something as best it can, just as a
plant grows,
or an animal seeks its food as best it can.
—CARL JUNG

Freud chose Carl Jung to be his most important colleague, and Jung enthusiastically learned more about the depths of the unconscious mind working with Freud.

In 1909 Freud's methods of uncovering unconscious drives and motivations were starting to get world-wide attention. Both he and Jung were invited to the US to deliver lectures. (This is covered quite interestingly in the 2011 movie "*A Dangerous Method*, directed by David Cronenberg.)

On the voyage, Jung and Freud each offered a personal dream to the other for interpretation.

But a rift was developing between the two; Jung felt Freud's insistence that dreams always were to be analyzed as repressions of sexual libido and fear of death was too restrictive. And Jung's keen interest in mysticism and spirituality irritated Freud.

It was what Jung referred to as his "big dream" which convinced him of a level of the unconscious beyond the personal that would ignite the angry split between them. And it would lead to Jung's insight of a "collective unconscious" which would in many ways create a doorway between western science and the spiritually expansive insights of ancient Eastern wisdom traditions. This "big dream" would eventually lead to a decades-long working relationship between Jung and a brilliant quantum physicist, both of whom were convinced that synchronicities uncovered a physical connection between the unconscious mind and external events.

Here is Jung's description of his "Big Dream:"

> *I was in a house that I did not know, which had two stories. It was "My House." I found myself in the upper story, where there was a kind of salon furnished with fine old pieces in Rococo style. On the walls hung a number of precious old paintings. I wondered that this should be my house. But then it occurred to me that I did not know what the lower floor looked like.*
>
> *Descending the stairs, I reached the ground floor. There everything was much older, and I realized that this part of the house must date from about the fifteenth or sixteenth century. The furnishings were medieval; the floors were of red brick. Everywhere it was rather dark. I went from one room to another, thinking, "now I really must explore the whole house."*
>
> *I came upon a heavy door, and opened it. Beyond it, I discovered layers of brick among the ordinary stone blocks, and chips of brick in the mortar. As soon as I saw this, I knew that the walls dated from Roman times. My interest was by now intense. I looked more closely at the floor. It was of stone slabs, and in one of these I discovered a ring.*
>
> *When I pulled it, the stone slab lifted, and I again I saw a stairway of narrow stone steps leading down into the depths. These, too, I descended, and entered a low cave cut into the rock. Thick dust lay on the floor, and in the dust were scattered bones and broken pottery, like remains of a primitive culture. I discovered two broken skulls, obviously very old and half disintegrated. Then I awoke.*[11]

Freud, interpreting the dream for Jung on their ocean voyage, went to his default position: dreams are repressed wish fulfillments from childhood rage against the parent of the opposite sex, the Oedipal and Electra Complexes. Jung wrote that he not only felt frustrated at Freud's limited insight about his dream, but at a deeper level, Freud's insistence every dream looks back in time to childhood repressions.

Jung's instinct, made even sharper by this "Big Dream," convinced him dreams not only look back, but provide important clues to what a person's psyche seeks to move towards in the future. He was never able to agree with Freud that the dream's meaning is maliciously hidden or withheld from consciousness. Instead, he believed that dreams are part of Nature and express things as best they can. He felt the house in his dream represented an image of the psyche, the ground floor standing for the first level of the unconscious. He noted that the deeper he went, the more alien and darker the scene became. In the cave, he discovered remains of a primitive culture, the primitive man within himself, a world scarcely reached or illuminated by consciousness.[12]

Freud may have been on the mark in citing the two broken skulls as representing Jung's desire to kill him as a father figure. This is what actually happened metaphorically—their sharp disagreement over this dream started the intractable conflict between the two, ending in a permanent split.

But while many of Freud's insights are still true and relevant today, history has shown Jung to be much more influential as a developer of depth psychology, the lens through which we can view much of our journey into **the Quantum & the Dream**.

Jung understood and wrote extensively that our dreams not only point back to repressed violent emotions, but, if we learn to recognize their patterns, offer deep insights into both our individual and collective urge for meaning and wholeness.

Looking at Jung's "Big Dream," some key features stand out:

- As with Einstein's exploration of the wave-particle duality in the quantum realm, Jung's opening description of the house in this dream reveals a deep paradox: He writes that it's "not a house I know," then follows that with the line, "It's my house." This confusion about his relationship to the house was the first inclination that dreams reveal details and meanings unique to the dreamer AND at the same time images generated by a much larger reservoir of meaning, the collective unconscious, shared by all human beings— inherited ways of seeing and acting in the world which created "the whole spiritual heritage of mankind's evolution, born anew in the brain structure of every individual."[13]

- This lack of clear distinction between the dream house being "not a house I know" and, at the same time, "my house" can be seen to mirror psychologically the wave-particle duality at the quantum level.

- In my experience doing dreamwork, it's a good strategy to start with the overall feel and movement of the dream before considering its particular content. Clearly in Jung's dream the overall feel and movement is a **"descending."** The dream starts from the upper story down to the ground floor, then down further through the "heavy door" to the steps leading even further down into the underground cave.

- Before Freud and Jung entered the world stage as hugely influential psychoanalysts, western civilization, influenced by the Abrahamic religions, had moved towards framing the spiritual journey of enlightenment as a rising up to heaven.

- But as the great mythologist Joseph Campbell pointed out, the great Western myths are filled with the descent into the underworld as part of "The Hero's Journey:" Orpheus descending to save his beloved Euridice; Theseus descending into the caverns to confront the Minotaur; Odysseus having to descend into Hades to encounter his dead comrades before reaching home; Dante's descent into Hell and Purgatory before arriving at Paradiso.

- Freud's publication of *The Interpretation of Dreams* brings this mythic understanding to modern science and psychology— the reality that enlightenment requires the willingness to descend into the dark unconscious. This literally creates the entirely new field of depth psychology. Jung's genius was to realize that if we are willing to explore our darkest dreams, they have within them not only the seeds of liberation from the powerful ties to the past, but in many instances, they expand our perception of both the present and potential futures.

- It's always worth paying attention when a dream repeats certain images. In Jung's dream, **bricks** are mentioned three times: The ground floor was made of "red brick" and opening the heavy door reveals "layers of brick" and "chips of bricks" cemented by mortar in the stone wall. Bricks are sturdy building materials, hardened through a heating process and cemented together in linear lines. They bind things together over long periods of time.

- As a metaphor for a way of thinking, these bricks can represent more of the left-hemisphere's preference for linear, clearly defined boundaries and a sturdy, well-defined structure, but at the cost of getting a "bigger picture," a more expansive context of what's happening to us in the dream. So the movement of Jung's dream from sturdy "layers of brick" to "chips of bricks" as he enters the mysterious staircase down to the cave beneath the house can be seen as a "chipping away" at the limited, linear, fixed kind of thinking which is incapable of receiving the deeper meanings of the dream. As Dr. Iain McGilchrist points out, ". . . only the right hemisphere can direct attention to what comes to us from the edges of awareness . . . novel experience induces change in the right hippocampus, not the left."[14]

- This movement of Jung's Big Dream from the structural integrity of the red brick ground floor, to the chips of brick in the mortar past the heavy door, to the disintegrating thick dust of the basement/cave can be seen as a calling to both break down the cemented psychological boundaries represented by brick and mortar—literally Jung's breaking down of the barriers Freud set up for interpreting dreams. And, from an even larger context, the erosion of science's certainty that the Universe operated under fixed and certain laws. This element of the dream connects to the insight we saw from science historian Thomas McFarlane at the beginning of this narrative: "In the 20th century the modern materialistic world view began to unravel in the face of scientific and psychological developments. It led a number of thinkers to consider that the human psyche may be more involved, in some mysterious way, with the observed properties of matter."[15]

- A key element of the dream is Jung's awareness within the dream that "now I really must explore the whole house," which reflects depth psychology's understanding that one of the greatest unconscious motivations of the human mind is to heal the sense of being fragmented and to seek Wholeness, a bigger picture of who we are and how we connect to the world at large.

• This instinct for a bigger picture, for "Wholeness," is implicit in the three key principles of quantum theory we will look at a bit further on in our journey. All three call for a much-needed shift in perspective from the left-hemisphere orientation of analyzing the world into separate parts to the right hemisphere intuition that any living system, including our inherent nature, can only be deeply understood by getting a sense of the "Whole," a willingness to seek greater depth of meaning which is always greater than the sum of its parts.

DINNER WITH CARL AND ALBERT

There is a synchronicity within the larger synchronicity of **the Quantum & the Dream:** In 1909, Einstein writes and publishes the first paper on the wave-particle duality of light which, as physicist John S. Rigden, says, could arguably be considered the start of quantum physics.

That same year, Jung shares with Freud his "Big Dream" of the house with three levels, all depicting different historic eras, and his descent through the mysterious doorway of eroded bricks, down into the dark cave underneath the foundation where he sees the two skulls in the disintegrated dust.

In 1909 both Einstein and Jung were still relatively unknown to the general public. Einstein's paper on the quantum theory of light was met with near total resistance from the scientific community—Jung's expanded perception of dreams was totally rejected by the master Freud, and therefore, not well received publicly.

By 1911 Jung's relationship with Freud was disintegrating (like those bricks in his dream). And it was in 1911 that Jung invites Albert Einstein to his home for dinner. (Both lived in the Zurich area of

Switzerland.) Fascinated by Einstein's mind-bending discoveries of relativity and wave-particle quantum paradox, despite not being able to follow the technical aspects of physics and mathematics, Jung would later write in a letter to a colleague, "It was Einstein who first started me off thinking about . . . time as well as space, and their psychic conditionality."[16]

These conversations with Einstein will lead to Jung's later forming a detailed theory of the collective unconscious first hinted at in his Big Dream, a repository of meaningful universal symbols he called "archetypes," an inherited group of patterns which reside underneath the personal unconscious and generate key images in the great myths, sacred stories, folk legends and teaching tales handed down through the ages. Among these key archetypes are the Mother, Divine Child, Sage, Warrior, Magician, King, Queen, Thief, Trickster, and Shadow. We'll meet a key Trickster archetype at the end of this section, one who points the way towards a deeper understanding of the psychological, philosophical, and spiritual dynamics needed to usher in the next renaissance.

In Jung's view, archetypes—the universal images percolating up from the collective unconscious—are "responsible for the organization of unconscious patterns...they have a "specific charge . . . a supernormal degree of luminosity."[17]

In addition to his first inklings of a collective unconscious, influenced first by Freud's emphasis on the psychological importance of dreams, then by Einstein's uncovering of how the subjective experience of an observer affects our sense of time in particular and external reality in general, Jung started to contemplate how archetypes might influence the nature of synchronicity, a doorway connecting the human psyche and the inner workings of the external world.

5

THE QUANTUM LEAP:
SCIENTISTS, *the* TAO, *and the* UPANISHADS

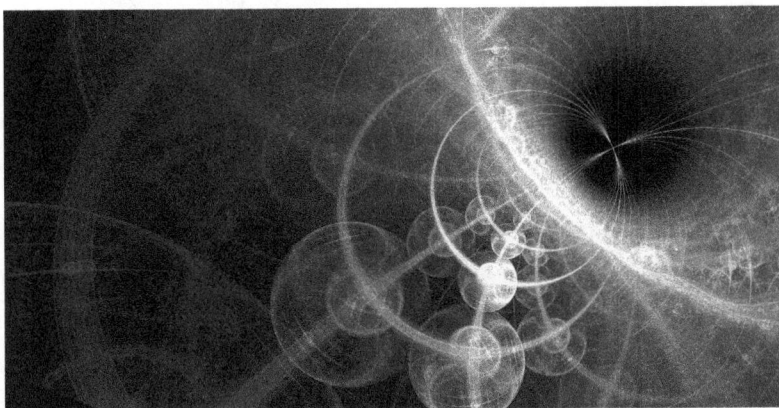

Tao lies beneath the reach of words. Yet if we
manage to enter the deeper region of our mind,
we can embark on a mystical descent to the
ineffable heart of being . . . .
—KAREN ARMSTRONG

What was being discovered in the descent into the quantum realm was looking more like the mystifying hall of mirrors in an amusement park or the surreal nature of a dream than the fixed, mechanical world view of Newton and Descartes which ushered in the Industrial Age.

After Max Planck set the wheels in motion with his accidental discovery of the quantum in 1900, the next few decades would find some of the brightest scientific minds on the planet not only willing to contemplate the deep paradoxes of the subatomic world, but actively engaging in philosophical and psychological conversations about what it all might mean. They questioned the fixed, long-held belief in science, philosophy, and psychology, still prominent today, that there is a clear distinction between mind and matter.

As brilliantly detailed in the seminal 1975 book *The Tao of Physics* by physicist/philosopher Fritjof Capra, many of the key players in quantum physics were inspired by various ancient Eastern spiritual insights. They each sensed a correlation between the dream-like visions of Taoist, Buddhist, Hindu, shamanic, and other mystical revelations emerging from their descent into the dark interior of the subatomic realm.

At the core of quantum physics, first revealed by Planck and later confirmed by multiple experiments, is the bizarre mystery that we saw Einstein address: At the subatomic level an entity such as a photon (which makes up light) or an electron (which makes up electricity) can appear to be a particle or a wave, depending on how the experiment is set up. This makes no logical sense.

Particles have finite, discernible boundaries and shoot around like billiard balls, whereas a wave is an underlying force of energy that spreads outwards. How could anything be neither definitively one or the other, but potentially either one?

Following a similar right-hemisphere "big picture" view as Einstein, Niels Bohr, a leading figure in quantum theory and future Nobel Laureate, came up with a philosophical response to the paradox of wave-particle duality which he called "complementarity."

Physicist David Harrison described Bohr's understanding of complementarity with this insight: we usually think of an electron as either a wave or a particle, but it is, in some sense, both particle and wave. Understanding Quantum Mechanics needs the ability to hold two viewpoints at once.

Here we have an intuitive connection between the ability "to think of two viewpoints at once" and one of the most enduring spiritual symbols, the yin/yang circle of the Tao.

When Bohr was honored with Danish knighthood in 1947, he had the opportunity to create a personal coat of arms. He chose as his central image this ancient yin-yang symbol of the Tao.

On the surface, the symbol appears to make a clear distinction between the light and the dark. But the division is not a straight line—it's a spiral, implying that the light and the dark are flowing in and out of each other. This is further amplified by the dark circle within the light section and a light circle within the dark section.

To get his mind around the paradox of quantum's wave-particle duality, Bohr adapted the underlying meaning of this enduring, ancient symbol: that the light and the darkness must be taken together for fuller understanding of the inner workings of the Universe, which is the very definition of complementarity.

From the perspective of **the Quantum & the Dream,** Bohr's philosophical concept of complementarity mirrors the advent of depth psychology begun by Freud and expanded by Jung—that to fully understand the human psyche requires both the dual nature of conscious awareness (light) and the creative, pervasive influence of the unconscious mind (darkness).

6

WELCOME *to the* QUANTUM FUN HOUSE:
NON-LOCALITY

*When we speak of physical reality, the very
concepts seem to cry out that one has to go
from one place to another; nothing can be
everywhere at once. This would require that
scientists would have to construct physics
completely from scratch, throwing out what
we thought was the essential nature of
reality . . . . But Heisenberg did just
this . . . backed by mathematical proof, he shows
that at the quantum level, there is no locality.
Until that is when something is measured.*
—DR. EVAN WALKER HARRIS

The second major player in the discovery of quantum theory is Werner
Heisenberg, who devoted his life to subatomic physics at the age of
twenty after attending a series of lectures given by Bohr. Heisenberg
enters the quantum spotlight for discovering the Uncertainty Principle
in 1925, a mathematical proof that one can measure the velocity of
a subatomic entity and one can measure its position—but one can't
know both at the same time. In other words, we can never know with
certainty where anything in the subatomic realm is located.

In 1927, while working with Bohr, he had a shocking but clear
realization about the limits of objective knowledge, since the very act

of observing affects the reality being observed. In fact, Heisenberg's Uncertainty Principle doesn't just "puncture" science's mechanized, clockwork universe—it evaporates it into thin air to reveal the mystical phenomenon of non-locality.

Non-locality, though confirmed experimentally, flies in the face of rational thought. It challenges the very foundation of objects being clearly separate from each other.

With the discovery of non-locality, the quantum world was looking more and more like a set of funhouse mirrors. Or a Zen koan. The universe at the quantum level has no specific "here," no "there," no specific "anywhere."

Again, the key underlying process here, reflecting **the Quantum & the Dream** synchronicity, is not just what Heisenberg discovered, but how he used primarily right-hemisphere contemplation to discover it. As detailed in the book *Helgoland: Making Sense of the Quantum Revolution,* by theoretical physicist Carlo Rovelli, Heisenberg traveled to the remote island of Helgoland in the North Sea so he could be alone without distraction in order to deeply meditate on the mysteries of quantum reality.

Just as was true of Niels Bohr, Heisenberg had an extensive interest in Eastern spiritual insights. Later in his life he claims to have experienced Kaivalya Samadhi, a Sanskrit term for a non- dualistic state of consciousness in which the observer becomes one with the observed.

Physicist Fritjof Capra met Heisenberg and later wrote that Heisenberg had spent time in India as the guest of celebrated Indian poet Rabindranath Tagore. Evidently the two men had long conversations about science and Indian philosophy, and Capra reported that Indian thought brought Heisenberg great comfort, and that he began to see relativity, interconnectedness, and impermanence as fundamental aspects of physical reality. These concepts, which had been so difficult for him and his fellow physicists, were at the basis of Indian spiritual traditions. In fact, Heisenberg said, "After these conversations with Tagore, some of the ideas that had seemed so crazy suddenly made much more sense. That was a great help for me."[18]

From the perspective of classical Newtonian physics and Cartesian philosophy, ruled by the laws of cause and effect and a clear distinction between mind and matter, left-hemisphere-dominant thinkers require empirical evidence using a sequential set of propositions as the basis of all truth. But the revelation that Uncertainty and non-locality were inherent in the very fabric of the Universe was a seismic blow to the materialist paradigm.

To the more right-hemisphere-oriented physicists and philosophers, this was a fanfare for creativity, intuitive freedom, and imagination. As quantum physicist Carlo Rovelli wrote, "I believe that one of the greatest mistakes made by human beings is to want certainties when trying to understand something. The search for knowledge is not nourished by certainty; it is nourished by a radical absence of certainty."[19]

We can connect the discovery of non-locality to the way our creative unconscious works when we are dreaming. Psychiatrist/ dream expert Dr. Montague Ullman, my first dream teacher, pointed out the correspondence between dreaming and the mysterious nature of non-locality in an address to the International Association for the Study of Dreams in 2006. There, Dr. Ullman stated: "We do strange things with time and space in our dreams as we impress them into metaphorical service. In the case of time the instantaneous condensation of past and present might be looked upon as a kind of subjective non-locality."[20]

Another observation on the dream-like reality of quantum non-locality from Philosopher David Albert:

> And so, the actual world is nonlocal. Period. We need to take seriously the idea that the world's history plays itself out not in the three-dimensional space

of our everyday experience or the four-dimensional spacetime of special relativity, but rather this gigantic and unfamiliar configuration space, out of which the illusion of three-dimensionality somehow emerges. Our three-dimensional idea of locality would need to be understood as emergent as well. The nonlocality of quantum physics might be our window into this deeper level of reality.[21]

It's said that "good things come in threes." The third key principle of quantum theory, Quantum Entanglement, might be the most mysterious and enlightening of them all.

THE MAIN ATTRACTION *at the* QUANTUM AMUSEMENT PARK: ENTANGLEMENT *or* "SPOOKY ACTION *at a* DISTANCE"

It's possible to link together two quantum particles — photons of light or atoms, for example — in a special way that makes them effectively two parts of the same entity. You can then separate them as far as you like, and a change in one is instantly reflected in the other. This odd, faster than light link, is a fundamental aspect of quantum science. Erwin Schrödinger, who came up with the name "entanglement" called it "the characteristic trait of quantum mechanics."

—PAUL COMSTOCK

Erwin Schrödinger developed the key equation for explaining the motion of non-material quantum waves. It's not the technical meaning of this equation which informs **the Quantum & the Dream**, but what Schrödinger revealed it implied: "Quantum Entanglement."

The word, "entanglement" here has a distinct meaning from the standard definition. In our waking life reality to be "entangled" is a restrictive, confining situation. But it's just the opposite when referred to at the quantum level.

If quantum phenomena can be so intricately connected that what happens to one quantum particle INSTANTANEOUSLY happens to the other, even if separated by billions of miles of space, what does this tell us about the nature of the universe we are part of?

Einstein, who had published the first paper on quantum theory, would not accept entanglement as a possibility—he famously referred to it as "spooky action at a distance." Even his expansive, imaginative mind couldn't tolerate it and he spent the rest of his life determined to disprove it.

While light travels faster than any other phenomena in the Universe—186,000 feet per second—it still takes some amount of time for light to travel billions of miles. So how can two entities be in communication *INSTANTANEOUSLY*?

How might we imagine this fascinating phenomenon psychologically? As psychiatrist/dream expert Dr. Montague Ullman observes, when we are dreaming, we've gone from the waking world of "discreet objects" into a domain where we are embedded within "resonant wave-like feelings seeking to embed themselves in symbolic imagery," a domain where we experience the "instantaneous condensation of past and present" and "the strange things we do with time and space in our dreams as we impress them into metaphorical service."

So, at the metaphorical/symbolic/dream level, we can make a case for some type of "instantaneous" connection, a valuable way to address the underlying process of our nightly dreams . . . but could this be the way the Universe works on a physical level?

A decade after Einstein died in 1955, physicist John Stewart Bell's Theorem revealed that Einstein's objections to quantum entanglement were not substantiated and that quantum theory was correct in predicting an inherently "connected" universe in which entangled particles could not be defined as being separated even if billions of miles apart.

An even more emphatic punch line was delivered a few decades later. In 1982 French physicist Alain Aspect was able to set up a physical experiment to show that under certain circumstances particles such as electrons are able to "communicate" instantaneously with each other regardless of the distance separating them. Quantum Entanglement is more than implied . . . it can be observed under certain laboratory conditions. (Note: In October, 2022, Alain Aspect and two other scientists who had demonstrated quantum entanglement in a laboratory received a Nobel Prize for this most significant empirical proof of quantum entanglement.)

We are all living in the intricately "entangled" connectedness of the quantum world. As an emergent quality of **the Quantum & the Dream** synchronicity, this phenomenon of quantum entanglement is worth contemplating more deeply, philosophically and psychologically.

We are used to thinking of objects separated by space as independent, but entanglement turns this idea on its head. One of the clearest and most talented communicators of the underlying implications of the quantum world, physicist Brian Greene, offers this view in his book, *The Fabric of the Cosmos*:

> But a class of experiments performed during the last couple of decades has shown that something we do over here (such as measuring certain properties of a particle) can be subtly entwined with something that happens over there (such as the outcome of measuring certain properties of another distant particle) without anything being sent from here to there. While intuitively baffling, this phenomenon fully conforms to the laws of quantum mechanics, and that was predicted using quantum mechanics long before the technology existed to do the experiment and observe, remarkably, that the prediction is correct.[22]

We can only imagine the reactions of Einstein, Jung, Bohr,

Heisenberg and Schrödinger to the 1982 laboratory results confirming Entanglement.

From our own vantage point, at a deeply intuitive level, this "entangled" phenomenon, or, as Greene writes, "Something that happens over here can be entwined with something that happens over there even if nothing travels from here to there—and even if there isn't enough time for anything, even light, to travel between them,"[23] strongly corresponds to the process underlying this entire journey— synchronicity. Specifically, the synchronicity revealed in the year 1900: the publication of Freud's *The Interpretation of Dreams* and Planck's unintended discovery of the quantum effect at the beginning and end of the first year of the Modern Age.

Those oriented much more to left-hemisphere perception—the majority of scientists and philosophers even to this day—wash the "entanglement" mystery away by proclaiming that while there are counter-intuitive phenomena existing at the quantum level of reality, these phenomena cease to have any effect at the practical, macroscopic levels of our daily lives.

The main resistance is based on the fact that quantum phenomena appear to only be shown experimentally at extremely cold temperatures, well below what a human brain could tolerate. But as the twenty-first century progresses, this "barrier" between quantum effects and the human brain is being challenged by noted scientists and philosophers.

For example, in the article "Do Quantum Effects Play a Role in Consciousness?" published in 2021 on the web site of *Physics World*, the magazine of the Institute of Physics, one of the largest physics organizations on the planet, the author writes:

> A quantum system – which might refer for example to the dynamics of a photon – is a delicate thing. Conventionally, quantum effects are observed

at low temperatures where this system is isolated from destructive interactions with its surrounding environment.

This would seem to exempt quantum effects from playing any role in the mess and fuss of living systems. . . . <u>However, this objection has to some extent been mitigated by research done in the broader field of quantum biology.</u> (Underlined emphasis mine.)[24]

It is becoming more evident that there could be a quantum effect even at the warmer temperature levels of brain waves and the unconscious. For example, it was confidently predicted that while immensely small sub-atomic entities such as photons, which make up light, and electrons, which create electricity, can be quantumly entangled, this could never be true for entities multiple times larger.

But in May of 2020, *Science Daily* reported that physicists successfully entangled an electrically charged atom and an electrically charged molecule, "showcasing a way to build hybrid quantum information systems that could manipulate, store and transmit different forms of data."[25]

On the deeply intuitive level, it makes no sense that, just because we aren't consciously aware of quantum effects, they have no relevance to how our minds operate. Should we be satisfied that quantum physics creates magnificent technologies but stop trying to understand its potential meaning and influence in the way we perceive ourselves and the world?

Let's jump back in time to the quantum physicist Erwin Schrödinger, who started the whole creative uproar when he surmised that his quantum equation implied "entanglement." As with Niels Bohr and Werner Heisenberg, Schrödinger was greatly inspired by ancient Vedic insights and wrote many statements which are clearly right-hemisphere oriented: "<u>The task is not to see what has never been seen before, but to think what has never been thought</u>

before about what you see every day." And "[Western science] is based on objectivation, whereby it has cut itself off from an adequate understanding of the subject of cognizance, of the mind."[26]

This "cognizance," the deeper knowing at the heart of the synchronicity generating this trip through **the Quantum & the Dream,** will be a significant factor in navigating the expanding speed of change and escalating anxiety here in the twenty-first century.

This does not refer to any specific belief: It's not about exclusively being a Freudian or Jungian, or exclusively accepting a Buddhist, Taoist, or Hindu vision, or buying into Bohr's Copenhagen interpretation of the quantum world. "Cognizance" is not just about what we consciously perceive—it goes deeper, to the expansive, intuitive connections pointed to at the threshold space between the conscious and unconscious mind, where signals percolate up into the right hemisphere of our brain, opening a doorway to a bigger picture of who we are and how we connect to the external world—not just the personal unconscious, but just as importantly, that deeper sense of the patterns of change emerging from the collective unconscious.

This right hemisphere, expansive "cognizance," this opening of the doorway between the known and unknown, the conscious and the unconscious, may determine humanity's ability to navigate the intensifying turbulence, uncertainty, complexity, and anxiety as the twenty-first century unfolds within the potential collapse of the very biosphere providing life on this planet as we know it.

8

PLUMBING *the* DEPTHS *of* ALCHEMY
&C the TAO

As Bohr's philosophical principle of Complementarity, Heisenberg's Uncertainty Principle and Schrödinger's quantum Entanglement were percolating through the world of physics, Carl Jung, deep into both his own dreams and those of his clients, receives a book on Taoist mystical healing titled *The Secret of the Golden Flower*.

The book is sent to him by his friend Richard Wilhelm who lived in China for twenty-five years and had become one of the leading translators of ancient Chinese philosophy. The book is a meditation on spiritual alchemy.

In the western material tradition, alchemy was the ancient attempt to transpose a base metal into gold, but in the ancient Taoist and gnostic mystical traditions, alchemy was practiced as a non-material spiritual purification, an inner trip where the limitations of self-centeredness can be transformed into clarity and the freedom inherent in the natural, spontaneous "original" mind.

Many left-hemisphere-oriented scientists (the majority) use Jung's attraction to alchemy and other mystical traditions to discredit his psychological theories. But in one of the great ironies of science history, it turns out that the scientific hero of the rational, mechanical nature of the universe, Sir Isaac Newton, was in fact immensely interested in and inspired by alchemy, but he kept this secret out of fear it would ruin his reputation as a scientific thinker.

Professor William R. Newman, Professor in the Department

of History and Philosophy of Science and Medicine at Indiana University, revealed Newton's alchemical inspiration:

> Newton employed alchemical themes in his physics, particularly in the area of optics. Newton's theory that white light is a mixture of unaltered spectral colors was bolstered by techniques of material analysis and synthesis that had a long prehistory in the domain of alchemy. But at the same time, he hoped to attain the grand secret that would make it possible to perform radical changes in matter.[27]

Centuries later, Carl Jung found alchemy, as understood in the ancient Taoist culture, an inspiration for his exploration of the unconscious mind. Based on work with patients and on his own inner explorations, Jung would come to see dream work and other descents inward as the alchemical transformation of the dark unconscious into the enlightened gold of expanded consciousness.

Psychoanalyst Dr. Murray Stein writes that Jung discovered a goldmine of psychological projections in the writings of ancient and medieval alchemists, and that their fantasies "revealed the structures and dynamics of the collective unconscious. . . . He was mining alchemical texts for the psychological gold hidden in their strange symbols and formulas."[28]

Among the parallels Jung saw between alchemy and dreamwork was how many of his patients found a liberating release from long-held traumas and anxieties by consciously working with him on their darkest nightmarish dreams—the transformation of dark, base emotional material into the light of conscious awareness.

Here, Jungian analyst Joseph Cambray identifies the correlation between ancient alchemy and the modern subatomic revelations of quantum physics: "Curiously, some cosmologies of the premodern era, such as the alchemical one, parallel that of subatomic physics with an original state prior to any differentiation of substances. They present a world of relations rather than objects, that is, attending to the interconnectedness of all things, where interactive processes appear more fundamental than discrete particles."

Spoiler Alert: We will later see how this "world of relations rather than objects" shows up in the twenty-first century more clearly as a potential scientific and psychological tool for global transformation. Just as Jungian analyst Cambray notes the "original state" as a "world of relations rather than objects," in 2020 noted quantum theorist Carlo Rovelli wrote, "We think of the world in terms of objects, things, entities. . . . To understand nature, we must focus on interactions rather than on isolated objects. . . . The world that we observe is continuously interacting. It is a dense web of interactions."[29]

9

THE PAUSE *that* REFRESHES...
RECAP *and* REFLECTIONS *on the* TRIP SO FAR

- Pivoting around the year 1900, the first year of the Modern Age, although unknown to each other, Dr. Sigmund Freud publishes *The Interpretation of Dreams* and physicist Max Planck discovers the quantum, initiating **the Quantum & the Dream** synchronicity in motion.

- In the year 1905 Albert Einstein, inspired by Planck's discovery of the quantum, a dream he had when he was eleven years old, and a creative thought experiment chasing a light beam, publishes his theory of relativity (time and space are connected to the subjective role of the observer) and the first quantum theory of light.

- A year later, Freud meets the younger psychoanalyst Carl Jung and they spend thirteen straight hours in excited conversation about dreams and the human unconscious; Jung seeks in his patients and his own inner work a deeper understanding of Freud's description of the unconscious, revealing its vast creative potential beyond the limited view as a turbulent reservoir of repressed sexual emotions and fear of death.

- Three years later, Jung presents his "Big Dream" to Freud. Convinced this was a call to see dreams as offering important information about the present and future, and to an expanded vision of dreams revealing a trans-personal, collective unconscious level, Jung is frustrated by Freud's insistence that dreams only look backwards. This is the start of what will become an irreparable split. It is important to note that just as Einstein did not prove Newton wrong, but only limited in regarding cause and effect as the sole principal dynamic of the universe, so Jung did not prove Freud wrong, only limited in his view of the dream as repressed emotions which look backwards into one's early upbringing. Freud deserves the credit for recognizing and making clear to a wider public the role of the dream as "the royal road to the unconscious" and understanding the correspondence between the unconscious mind and the deep psychological truths inherent in ancient Greek myths.

- That same year, 1909, Einstein publicly expresses his view that to understand the nature of light, one has to accept the paradox of wave-particle duality.

- In 1911, Jung, having become aware of Einstein's boundary-expanding discoveries, invites him to his home for a series of dinners. Jung later claims that while he could not keep up with the math and physics, his intuition about relativity's boundary-expanding perception of time and space and quantum's wave-particle paradox, combined with his interpretation of his "Big Dream," lead him to both a deeper understanding of the personal unconscious and the reality of a much larger "collective unconscious."

- Beginning in 1913 with Niels Bohr's explanation of the interior of the atom which would win him a Nobel Prize, and continuing for more than a decade through collaboration with other brilliant scientific minds such as Heisenberg and Schrödinger, the previously unknown creative interior of the quantum world emerges to reveal three mystifying implications about the nature of reality: wave-particle duality, the Uncertainty Principle, and Quantum Entanglement. At both the quantum level and the unconscious level of the human mind, the long-held paradigm in Western culture of rational boundaries among past, present, and future and the mechanized, clockwork perception of the universe, like those disintegrating bricks in Jung's dream house, are breaking down.

- On a parallel wavelength, as Einstein, Heisenberg, Bohr, Schrödinger, and their scientific colleagues descend further into the interior of the subatomic quantum realm, Freud and Jung present a new psychological model, depth psychology, including methods for descending deeper into the interior of our psyches, uncovering the latent meaning of our dreams (which modern neuroscience show are more closely connected to the right hemisphere of our brains), widening the doorway through which our psychological needs and desires for meaning percolate up from the dark unconscious to the light of conscious awareness.

10

BRIEF SIDE TRIP:
THE SECRET THEATER
of SILENT MONOLOGUE

*It begins in wonder, intuition, ambiguity,
puzzlement . . . it progresses through being
unpacked, inspected from all angles and
wrestled into linearity by the left hemisphere;
but its endpoint is to see that the very business
of language and linearity must themselves be
transcended, and once more left behind. The
progression is familiar: from right hemisphere, to
left hemisphere, to right hemisphere again.*
—IAIN MCGILCHRIST

When we shift from tightly focused left-hemisphere thinking to a more expansive big-picture context, we take boundary-shaking, totally unpredictable leaps into the unknown capabilities of right-hemisphere thinking. This process of shifting to a more right-hemisphere-oriented awareness, this contemplative "flow" state of mind—which Einstein, Jung, Bohr, Heisenberg, Schrödinger, and a fourth world-renowned quantum theorist we are about to meet, known as "the scourge of god," used to uncover their amazing discoveries about the nature of the universe and the nature of our unconscious mind—is available to all of us.

To make this shift, however, requires a willingness to delve into the dark, ambiguous, puzzling, paradoxical, yet often playful, potentially transformative nature of the unconscious mind.

Modern brain imaging technologies such as MRI (magnetic resonance imaging) and EEG (electroencephalogram), both of which were developed using quantum physics, allows us to literally see the connection between the creative unconscious and the right-hemisphere region of our brains.

As previously noted, Dr. Iain McGilchrist explains that:

> The major difference between the hemispheres lies in their relationship with the unconscious mind, whether that means the dream state . . . or what we experience or bear in mind without being aware of it. Whatever does not lie in the center of the attentional field, where we are focused, is better yielded by the right hemisphere, and the left hemisphere can sometimes show surprising ignorance of it.[30]

The gateway threshold between our dreams and the right hemisphere is now well documented, as scientists have proven that during REM sleep and dreaming, blood flow to the right hemisphere is greatly increased. EEG data also indicate the predominance of the right hemisphere in dreaming. For us to get a deeper understanding of our dreams and intuitive visions requires a shift in consciousness to the form of awareness given to us by evolution that is wired for transcendence and creative mystery.

Julian Jaynes, a psychological researcher at both Princeton and Yale, points out the primary alignment of the right hemisphere in first receiving and then in playing with and relating to the thoughts and feelings that percolate up from the unconscious mind. He offers a wonderfully provocative description of the unconscious as a "secret theater of speechless monologue and prevenient counsel, an invisible mansion of all moods, musings, and mysteries. . . . An introcosm that is more myself than anything I can find in a mirror."[31]

Pioneering neuroscientist Benjamin Libet, by hooking up volunteers to record their brain waves, showed that our so-called conscious decisions are in fact preceded a split second earlier in the brain before any conscious activity shows up on display screens. While Libet made clear he felt strongly that we humans have free will, his experimental results reveal the unconscious to be the initiator of the vast majority of our thoughts, feelings, beliefs, and actions, and that consciousness is the top layer of a much deeper, more pervasive resource.

With this understanding in mind, McGilchrist described the relationship between the left and right hemispheres of our brain as it deals with the mysterious, sometimes frightening, but potentially creative/healing thoughts and feelings which come into awareness from the deep unconscious, in the epigraph at the beginning of this chapter. It is quite useful that the left hemisphere of our brains analyzes and clarifies some of the ambiguity and puzzlement of unconscious messages first arriving in the right-hemisphere. But one of McGilchrist's key insights is that the left hemisphere is not wired to share, but to grasp and hold onto certainty and control. It is our conscious responsibility to shift back to right hemisphere perception, to maintain the "wonder, intuition, ambiguity and puzzlement" which seeds our lives with transcendent awe and the ability to see the world anew.

11

THE "SCOURGE *of* GOD" FINDS JUNG

It would be most satisfactory if physics and
psyche could be seen as complementary aspects of
the same reality.
—WOLFGANG PAULI

The orientation at the beginning of this trip opened with a description of *synchronicity*: "a falling together in time," two or more events which are clearly not a result of cause and effect, but too meaningful to be brushed away as mere coincidence.

The fourth leading quantum physicist we will look at was so intrigued by synchronicities and by the healing gained from working with his emotionally charged dreams that, beginning in the early 1930s, he initiated what will become an over-twenty-year relationship with the most prominent depth psychologist of the time, Carl Jung.

Together they contemplate and seek to discover a scientific/ psychological connection between the inner psyche of the mind and the mysterious quantum revelations of wave-particle duality, Uncertainty, and Entanglement.

Born in 1900, the same year Planck discovers the quantum and Freud publishes *The Interpretation of Dreams*, Wolfgang Pauli becomes one of the most prominent theorists of quantum physics as well as an emotionally fractured personality whose personal life was falling apart.

Contemporary physicist F. David Peat notes that, at the age of twenty, Pauli had written a two-hundred-page article on the theory of relativity that was praised by Einstein, who was impressed by Pauli's youth and the depth of psychological understanding, as well as his profound insight.[32]

Of the qualities Einstein saw in Pauli, the one of most interest to **the Quantum & the Dream**, is the psychological understanding of the development of ideas. For it would be this search into his own emotional issues that would inspire Pauli to seek a doorway between the quantum world and the unconscious mind.

Despite his rising fame as a quantum physicist collaborating with Bohr and Heisenberg in the 1920s, Pauli was falling apart emotionally. The breakup of his marriage occurred just a few years after his father left his mother and she committed suicide.

During the day he was earning a professional reputation as a brilliant professor of physics, although his difficult personality and heretical beliefs gave him the nickname, "the Scourge of God." At night he was frequenting Hamburg, Germany's underworld of prostitutes, drugs, and alcohol.

Pauli's father recommended he seek the help of psychoanalyst Carl Jung. This led to the decades-long relationship which included deep analysis of dozens of Pauli's dreams and the mutual fascination of the two with the phenomenon of synchronicity.

In his essay on Pauli, F. David Peat describes Pauli's psychological progress as opening a dialogue with the deepest levels of his unconscious mind, which then began to teach him. These encounters with the unconscious culminated in a vision—the World Clock—so sublimely harmonious that it produced an experience akin to religious conversion. In the dream, rotating discs expressed the mysterious harmony of the cosmos, and its symbolism united two worlds. This central theme of the unification of two worlds occurred again and again in both Pauli's waking and dreaming life.

Peat goes on to describe Pauli's other dreams in which he is visited by an "exotic woman." Pauli believed she was his soul, and began to see that the most important issue was the modern scientific conception of the world's lack of soul. He was driven by a vision of returning the soul to the world.[33]

A note on this "exotic woman" who showed up in Pauli's dream, whom he considered to be his soul: following Freud's discovery of the connection between ancient Greek myths and the unconscious mind, the "exotic woman" of Pauli's dream can be seen as connected to the myth of Psyche, the ancient Greek Goddess of the Soul. It is from her we get the word, "psychology." As previously noted, the word, "psyche," comes from the Greek word which can be translated as both mind and soul.

Depth psychology, which Freud created and Jung greatly expanded, studies the whole mind, conscious and unconscious, and is therefore much more reflective of the nature of "psyche," which is both the rational, visible mind celebrated in the Newtonian/Cartesian paradigm AND the much larger, deeper, darker, more influential depths where Psyche and the mysteries underlying quantum theory reside.

According to scientist/author Arthur I. Miller, Jung and Pauli sat in Jung's gothic mansion on the shores of Lake Zurich for hours on end, drinking vintage wines, dining on fine food, and smoking the finest cigars, while their discussions ranged from physics and whether there is a cosmic number at the root of the universe, to ESP, UFOs, Jesus, Yahweh, Armageddon, and Pauli's dreams. Jung observed that Pauli was able to inhabit the "darkest hunting ground of our time": the no-man's-land between the psychology of the unconscious and physics.

Depth psychology promotes the enormous, rich, soulful trip inward, a trip requiring a willingness to explore the dark interior of our being—the essential nature which modern consumer culture discourages, encouraging us to remain scared of the "dark" and the "creative unknown."

This is what Carl Jung was addressing with one of his most provocative insights:

"A secret unrest gnaws at the root of our being. Dealing with the unconscious has become a question of life for us."

The literary and cultural critic Maria Popova, commenting on Jung and Pauli's relationship, observed that their conversations and correspondence explored fundamental questions about the nature of reality, using the dual lens of physics and psychology, each using the tools of their particular expertise to "shift the shoreline between the known and the unknown." Together, they found common ground in the analogy of the atom, with its electrons orbiting around the nucleus, and the self, with its central conscious ego and its ambient unconscious.

Jung wrote to Pauli that the nucleus of the atom "... is an excellent symbol for the source of energy of the collective unconscious, the ultimate external stratum of which appears in individual consciousness. As a symbol, it indicates that consciousness does not grow out of any activity that is inherent to it; rather, it is constantly being produced by an energy that comes from the depths of the unconscious and has thus been depicted in the form of rays since time immemorial."[34]

In her article for *New Scientist Magazine*, Amanda Gefter proposes that this observation of Jung's mirrors the philosophical insight of Niels Bohr's principle of complementarity. She writes that both were entranced by Bohr's notion, which "asserts a reconciliation of opposites, such as wave and particle, in a framework not unlike that of yin and yang. Jung learned that a basic principle of alchemy – reconciliation of opposites into a unity – pervades quantum physics, too."[35]

As we saw earlier, according to Bohr, to understand the quantum level of reality one must perceive sub-atomic entities as particles in some instances, waves in others: both perceptions are needed to get a full understanding. According to Jung, to understand the unconscious one must perceive it both as a unique reflection of the individual mind in some instances, and the reflection of a collective level inherent in all humanity in others. This insight will have a significant impact when we meet up with **the Quantum & the Dream** as it emerges in the transition to the twenty-first century.

Jung and Pauli were never able to come up with a satisfactory explanation of synchronicity and the vast majority of the scientific world gave little if any credence to their explorations. In fact, a prominent mathematician, on reviewing the Jung/Pauli correspondence believed they were both completely mad.

From the left-hemisphere's need for certainty and its inability to consider the creative ambiguity of paradox, Jung and Pauli's exploration would appear "mad." But it takes a little madness, if by madness we mean a willingness to challenge the entrenched materialist paradigm, which both quantum theory and depth psychology were clearly doing. Jung and Pauli's collaboration speaks to the awareness that it's just as "mad" to ignore the vast realm of deep unconscious information which, as modern neurological evidence has shown, is responsible for some 95 to 99 percent of our perceptions, feelings, and ideas about who we are and how we connect to the outside world.

Interestingly, decades later, in 1999, Fritjof Capra, author of the best-selling book, *The Tao of Physics: An Exploration of the Parallels between Modern Physics and Eastern Mysticism*, confirmed the relevancy of the Pauli/Jung exploration, saying that the recently-emerged concept of interconnectedness underlines the similarities between the physicist's view and that of mystics. Intriguingly, it also raises the possibility of Jungian psychology's relationship to sub-atomic physics.

IS TIME MELTING AWAY?
DALI *and* SYNCHRONICITY
WITHIN SYNCHRONICITY

In 1931, less than a year before the Nobel-Prize-winning quantum physicist Wolfgang Pauli seeks out Jung for help in understanding his disturbing dreams, the surrealist painter Salvador Dali paints and exhibits what will become the twentieth century's most famous dream image of avant-garde art: those melting pocket watches Dali titled "The Persistence of Memory."

These iconic melting watches are directly connected to **the Quantum & the Dream,** for according to an essay on the website of the Dali Museum, Dali's connection to Freud is well-documented, starting with *The Interpretation of Dreams,* which was, according to the artist, one of the capital discoveries of his life. He was well-versed in Freud's evaluation of the unconscious for surreal and artistic inspiration, and regarded dreams and imaginings as central to human thought. He finally met Freud in 1938, and Dali's portrait of Freud hangs in Freud's last home. Dali used Freud's ideas about dream analysis as a source for his paintings, with their combination of precise realism and dream-like imagery.[36]

In addition to the dream, Dali was fascinated and inspired by both the theory of Relativity and Quantum theory. As artist and art educator Carmen Ruiz writes, "What do Stephen Hawking, Albert Einstein, Sigmund Freud, "Cosmic Glue," Werner Heisenberg, Watson and Crick . . . and Erwin Schrödinger have in common? The answer is . . . Salvador Dalí."[37]

Dali had dozens of books in his library on quantum physics and evolution which informed his surrealistic visions, including those mysteriously melting pocket watches.

Dali never answered the often-asked question, "What is the meaning of those melting time pieces?" As with our dreams, there are many potential meanings. In the instance of an enduring work of art such as "The Persistence of Memory," those melting time pieces are a powerful example of what Jung called an "archetype," universal images rising up from the collective unconscious which he described as "responsible for the organization of unconscious psychic processes . . . they have a 'specific charge'... a supernormal degree of luminosity."[38]

We can connect Dali's iconic melting time pieces to the description offered earlier of Heisenberg's Uncertainty Principle, which "punctured the centuries-old, firmly held belief that the universe and everything in it operates like clockwork."[39]

And we can connect those melting pocket watches with the insight Einstein introduced at the beginning **The Quantum & the Dream**: "The distinction between past, present and future is only a stubbornly persistent illusion."

Then there's the intriguing question: Why did Dali choose to call this depiction of time melting away as "The Persistence of Memory?" What is it that "persists" if the classical notion of time and of the distinction among past, present, and future melt away?

Dali Universe, a team of art historians which exhibits Salvador Dali's artwork worldwide, say that Dali's message has to do with the subliminal unconscious present in our daily lives that has more power over us than do the man-made objects of the conscious world. In Dali's painting, time has no importance and objects made by humans, such as the watches, are transitory.[40]

It's interesting to note that in October of 2021 it was reported that the most popular new digital Emoji permeating the World Wide Web was "The Melting Emoji," a face melting away even as it displays a faint smile, reflecting the current mood of escalating anxiety caused by climate change devastation, and the feeling that many of the expected norms of daily life are "melting" away.

Following this synchronistic pattern, Dali's iconic melting pocket watches grew out of the Surrealist Movement in art and literature which first appeared in 1924, the same year Niels Bohr, Werner Heisenberg, and Wolfgang Pauli were meeting and corresponding about the paradox of wave-particle duality.

Surrealism took the culture by storm in the wake of World War I, changing the face of modern art and influencing philosophy and social movements as well.

The Surrealist Movement overall, as with Dali individually, was in large part a reaction to the new discoveries about the unconscious starting in the year 1900 with the publication of *The Interpretation of Dreams*. The founder of the Surrealist Movement, poet and artist André Breton wrote, "Surrealism is based on the belief in the omnipotence of dreams, in the undirected play of thought."[41]

Interesting phrase, **"the undirected play of thought."** Who or What is the director?

Travelling forward from 1924 to 1931–1932, and taking a more global viewpoint, we can observe another synchronicity: not only did Wolfgang Pauli, one of the founders of quantum theory, start exploring the depths of his dreams with Carl Jung less than a year after Dali's melting pocket watches showed up in an exhibition, but we have the "timely" connection to Pauli's dream of a harmonious World Clock.

While I could not find an image of the World Clock from Pauli's dream, we have physicist F. David Peat's description of its huge rotating discs as recounted by Jung:

> The two disks belong to the two universes of the conscious and the unconscious.... The whole figure together with its elaborate internal movement

is therefore a mandala of the Self, which is at one and the same time the center and the periphery of the world clock. In addition, the dream could also stand as a model of the universe itself and the nature of space-time.[42]

Contemplating Pauli's dream with its two rotating discs of a world clock which generate a sense of Unity at the core of the Universe, along with Dali's surreal dream-like image of melting pocket watches, can lead to the insight that pocket watches are only used by individuals whereas the "unifying" symbol of a world clock is a global, collective archetype.

One potential meaning to be drawn from these two "timepiece" dream images, one from the right-hemisphere vision of an artist, the other from the right-hemisphere vision of a quantum theorist, both emerging from a fascination with the unconscious, is the benefit of contemplating the connected gateway between the unique individual mind and that of the collective unconscious.

13

"SHUT UP *and* CALCULATE":
THE LEFT HEMISPHERE PUSHES BACK

The fascinating, revelatory, paradigm-shaking insights emerging out of the minds of Freud, Einstein, Jung, Bohr, Heisenberg, Schrödinger, Pauli, and Dali, are repressed and virtually pushed out of view in the mid-to-late 1930s.

Tired of being challenged by the complex mysteries of quantum theory and the volatile reality of the creative unconscious, the materialistic mind-set which rose up was perfectly captured in the phrase attributed to contemporary physicist at Cornell, David Mermin: **"Shut up and calculate."**

One of the main reasons for the loud voice demanding a refocus on left-hemisphere, practical results was the mandate of the US government and its allies at that time, given the concern of rising fascism in Europe, to use knowledge of quantum physics to build the first atomic bomb.

But even after the war, the majority of quantum physicists continued to focus on the practical, technological potential at the total expense of the exploration of meaning and expanded vision. This was echoed in psychology under the heavy thumb of behavioralism, and in philosophy under the thumb of logical empiricism, rejecting any theory or insight not provable by certain, empirical evidence. Virtually drowned out were the creative musings, imaginative leaps and "undirected play of thought" which infused the perceptions of Einstein, Bohr, Heisenberg, Schrödinger, Pauli, and Jung.

This call to **"shut up and calculate"** is a common voice from the left-hemisphere of the human brain, capable of intense focus on specific goals and creating brilliant new technologies of the modern age, including semiconductors, silicon chips, lasers, fiber optics, MRI medical breakthroughs, the super computer, the smartphone and the Internet—a spectacular track record.

But the unseen consequences?

Where is the search for Meaning? Where is the love of wisdom, of the sense of wholeness which comes from exploring the deep interior of our psyche, of Einstein's insight that **"imagination is more important than knowledge?"**

Where is the balance needed to perceive and understand the devastation which awaits all humanity and virtually all living species on Earth here in the twenty-first century? It is the drastically out of balance, materialistic-grasping, technologically-obsessed mindset that gives rise to the irresponsible toxicity threatening the very biosphere responsible for maintaining life on this planet.

An interesting note: At the first successful testing of the atom bomb at Los Alamos, the lead scientist, J. Robert Oppenheimer, watching the massive burst of energy and toxic cloud rising into the atmosphere, quoted aloud the words of Shiva, the ancient Hindu god: "Now I am become Death, the Destroyer of Worlds." (Synchronistically, we will meet the powerful archetype of Shiva, creator and destroyer of the Universe, in a significantly different context, when **the Quantum & the Dream** moves towards and then into the twenty-first century.)

From the mid-1930s until their deaths, Einstein, Bohr, Heisenberg, Schrödinger, and Pauli continued to correspond, write articles, and give talks offering philosophical/spiritual insights into the potential meanings of quantum theory specifically, and into core philosophical life issues.

Jung and Pauli continued to meet and correspond about the potential bridge between quantum physics and the unconscious mind. But these important voices were barely heard outside small circles.

As for the mindset of "shut up and calculate," Dr. Iain McGilchrist, the philosopher and neuroscientist quoted often in this section, gives us a more expansive, panoramic, right-hemisphere-oriented view of our minds in operation. He states that for human beings there are two fundamentally opposed realities, the left and right hemispheres, two different modes of experience. Each is important in bringing about the recognizably human world, and that the bi-hemispheric structure of the brain is at the root of this difference. Although the hemispheres clearly need to cooperate, he believes they are actually in a sort of power struggle, which explains many aspects of contemporary Western culture.[43]

No matter how hard the culture in general or we as individuals pressure our conscious minds to repress it, the unconscious, known to be the inherent force behind 95 to 99 percent of our thoughts, beliefs, perceptions, and actions eventually has its say.

Carl Jung put it right out there: "The psychological rule says that when an inner situation is not made conscious, it happens outside as fate."[44]

14

"OPEN UP *and* CONTEMPLATE": THE END *&* THE BEGINNING

This World is not Conclusion.
A Species stands beyond –
Invisible, as Music –
But positive, as Sound. . . .
—EMILY DICKINSON

Sigmund Freud, whose publication of *The Interpretation of Dreams* shines a light on the "royal road to the unconscious," and sends seismic, creative waves of expanded perception through psychology, philosophy, science, art, music, literature, and the culture at large, dies in 1939.

Max Planck, whose unintended discovery of the quantum only months from the publication of *The Interpretation of Dreams*, and which unleashes enlightened explorations into the interior of the subatomic domain, creating quantum physics—the most successful scientific theory of all time—challenging the very meaning of the relationship between observer and the observed, dies in 1947. (Note: As previously described, at the time of his discovery of the quantum Planck was an entrenched left-hemisphere-oriented rationalist who was more bothered by the inherent paradox of the quantum than excited by it. Yet, as a great example of shifting from the left hemisphere to the right hemisphere, after two decades of working with quantum colleagues and contemplating his discovery more deeply, stated: "Science enhances the moral value of life, because it furthers a love of truth and reverence ... reverence, because every advance in knowledge brings us face to face with the mystery of our own being.")

Albert Einstein, who said his entire career had been a meditation on a dream he had when he was eleven years old, who generates a dynamic paradigm-shift in the understanding of the relationship between observer and the observed, and who was named "Person of the Century" by *TIME* magazine, dies in 1955.

Wolfgang Pauli, Nobel Prize winner for quantum physics, irascible critic known as "The Scourge of God," who for decades sought out the deeper meaning of his dreams and spent over two decades in conversation with Carl Jung about the nature of synchronicity, dies in 1955.

Carl Jung, who first learned at the side of Freud, then expanded our vision of the unconscious as a creative source of transformation and uncovered a deeper layer, the collective unconscious, which connects all of humanity past, present and future, dies in 1961.

Erwin Schrödinger, who developed the most effective quantum equation, opened the doorway to the mystical phenomenon of "Entanglement," who once said "The task is not to see what has never been seen before, but to think what has never been thought before about what you see every day," and was dedicated to the spiritual practice of Advaita Vedanta, dies in 1961.

Niels Bohr, often referred to as the "Father of Quantum Physics," who chose the spiritual/mystical yin/yang symbol of the Tao for his personal coat of arms and once said "How wonderful that we have met with a paradox. Now we have some hope of making progress," dies in 1962.

Werner Heisenberg, who uncovered the Uncertainty Principle and Non-Locality and was deeply inspired and influenced by ancient Hindu mysticism, dies in 1976.

All of these brilliant minds, either well-schooled in traditional studies of math and science, or in the case of Freud and Jung, trained as medical doctors, shifted to more right-hemisphere wavelengths, the descent into the laboratory of the inner mind, using the tools of dreams, thought experiments, meditation, spiritual insights, and imaginative leaps emerging out of the dark, creative unconscious.

But as noted, by the mid-to-late 1930s, the left-hemisphere shout of "shut up and calculate" rang loudly through the second half of the twentieth century: quantum physicists, for the most part, shunned the search for philosophical and psychological meaning as a distraction from the need to focus on practical results and better technology.

Logical positivism took control of contemporary philosophy, insisting that if it can't be empirically demonstrated in the outside world, then it has no important meaning, forgetting the root mission of philosophy: the "love of wisdom" which cannot be contained within the box of certainty and logic.

Behaviorism took control of most of psychology: If it can't be objectively observed and measured, it isn't relevant; human reactions are the results of responses to outside stimuli, and the inner life isn't worth studying.

We can imagine the right-hemisphere response to "shut up and calculate":

"Open Up and Contemplate."

As contemporary physicist Arthur Zajonc writes in his book, *Catching the Light: The Entwined History of Light and Mind,* "No matter how brilliant the day, if we lack the formative, artistic power of imagination, we become blind, both figuratively and literally. We need a light within as well as daylight without for vision: poetic or scientific, sublime or common."[45]

He added that the mind is subtly and usually unconsciously active in our sense of sight, forming and reforming the world we see, and so we participate in sight.

This descent into the unconscious realm of the human mind, and the retrieval of creative insights percolating into the right hemisphere of the brain for conscious attention, is the continuing clarion call of **the Quantum & the Dream.**

And we will observe its deep waves of influence into the latter stages of the twentieth century, crossing into the new millennium, providing insights to help us navigate the accelerating turbulence of cataclysmic change.

Revisiting the description by philosopher/scientist Thomas McFarlane: "In the 20th century the modern materialistic world view began to unravel in the face of scientific and psychological developments. It led a number of thinkers to consider that the human psyche may be more involved, in some mysterious way, with the observed properties of matter."[46]

No matter how loud the voice of "shut up and calculate" from the left hemisphere, it can't hide this "unraveling" of the materialist view as to who we are and what brings deep meaning to life.

MOMENT OF REFLECTION: WIDENING THE APERTURE OF THE UNCONSCIOUS MIND'S INNER CAMERA

If we take a moment to contemplate **the Quantum & the Dream** synchronicity so far:

In the year 1900, the first year of the modern age, we can see the pivotal nature of "the descent." Sigmund Freud publishes *The Interpretation of Dreams* and Max Planck unexpectedly discovers the quantum, initiating two significant, parallel paths:

> *the descent into the creative darkness of the human psyche and*
> *the descent into the mysterious darkness of the sub-atomic realm,*

both of which, as personified by Einstein, Jung, Bohr, Heisenberg, Schrödinger, and Pauli, revealed a dramatic shift from the clearly defined focus of the left hemisphere of the brain to the more expansive synaptic jump cuts, imaginative leaps, and open-minded, "big picture" capacities of the right hemisphere, challenging the step-by-step, rational, materialistic Newtonian/Cartesian mindset, a world view which has dominated Western culture for hundreds of years.

SPEEDING TOWARDS *the* NEW MILLENNIUM: SIGNALS *from the* RIGHT HEMISPHERE

I realized that the story of ourselves as told by science—our cosmology, our religion—was incomplete and likely flawed. I recognized that the Newtonian idea of separate, independent, discrete things in the universe wasn't a fully accurate description. What was needed was a new story.

—ASTRONAUT EDGAR MITCHELL

In the 1960s, 70s, and 80s, there were scientific philosophical, and spiritual voices which occasionally could be heard through the din of "shut up and calculate" and the urge to repress the unraveling of the materialist world view.

- In the 1960s and 70s Marshall McLuhan, a literary professor who shifted his attention to the underlying, primarily unconscious patterns of media and their effect on our thinking, became one of the most sought-after speakers and consultants in the world. Often referred to as the "Prophet of the Electronic Age," McLuhan coined the phrases "Global Village" and "the Medium is the Message" to express the relative shrinking of time and space in the modern world of instantaneous electronic communication.

Known for using riddles, jokes, obscure references, and koan-like phrases to wake up his audiences to the huge shift in consciousness caused by the development of electronic communication, McLuhan once answered a TV interviewer's question, "Why are you so difficult to understand?" with: "Because I use the right hemisphere when they're trying to use the left."[47]

We'll delve into McLuhan's insights and their relevance to the New Renaissance more deeply later on in our trip through **the Quantum & the Dream,** when we explore the shift from the printed page to the digital screen.

- In 1962 marine biologist/conservationist Rachel Carson publishes *Silent Spring,* which burst onto the cultural landscape decrying the pervasive use of toxic chemicals and their lethal consequences on human health and the overall health of the planet. Attacked both for her audacity at challenging the powerful chemical lobby and for being a strong-willed woman scientist (she was attacked in public for being "hysterical" and "irrational") her meticulous research and a talent for strong narrative forced the elimination of the commercially profitable chemical DDT and is credited with single-handedly igniting the modern environmental movement.

- In 1971, Edgar Mitchell (sixth astronaut to walk on the moon), after an overwhelming spiritual experience on the trip home sparked by the vision of Earth, sun, and moon from outer space, forms the Institute for Noetic Sciences, whose mission statement is "to use scientific exploration and personal discovery to push beyond the current limits of human knowledge."

- In the early 1970s, evolutionary biologist Lynn Margulis publishes books and articles challenging the materialist core of Darwin's theory of evolution, citing evidence that, from the cellular structure of micro-organisms to galaxies, there is a deep pattern of collaboration referred to scientifically as "symbiosis." In 1974 she co-authors an article presenting the "Gaia Hypothesis," proposing evidence that the Earth is a living system, not merely a material planet. A woman more than willing to push through the prejudice of entrenched male power in academia, she earned the nicknames "science's unruly earth mother," and "vindicated heretic." (More on Margulis' important role in the Third Shift of our trip.)

- In 1975 physicist Fritjof Capra publishes *The Tao of Physics: An Exploration of the Parallels Between Modern Physics and Eastern Mysticism*. Considered controversial, the book became a surprise international best-seller.

Just before publication, knowing his book would be attacked by mainstream scientists, Capra reviewed the key insights of the book with none other than key quantum theory founder Werner Heisenberg, then in his mid-seventies. Capra writes that he told Heisenberg that he saw two basic themes running through modern physics theory, which were also the themes of all mystical traditions: the fundamental interdependence and interrelatedness of all phenomena. At the end of his long presentation, Heisenberg said simply that he was in complete agreement.[48]

The Tao of Physics was written in response to an epiphany Capra had a few years earlier. It was a late summer afternoon and Capra sat on a California beach, feeling his breathing and watching the waves, when he suddenly became aware of everything—sand, rocks, water, air—as taking part in a gigantic cosmic dance. What he had learned through graphs, diagrams, and mathematical theories about high-energy physics came to life as he saw and heard the Dance of Shiva, the Hindu god responsible for the creation and destruction of the Universe. Capra, who had been trained in detailed analytical thinking, was so overwhelmed that he burst into tears.[49]

The Tao of Physics has sold over one million copies in forty editions around the globe. Capra updated the book for its twenty-fifth anniversary in the first month of the new millennium, January, 2000.

(Note: This is the second reference on our trip through **the Quantum & the Dream** to Shiva, the archetypal Creator and Destroyer of the Universe emerging out of Hindu mythology, the first reference being quantum physicist J. Robert Oppenheimer upon witnessing the massive explosion of the first test detonation of an atomic bomb. We will witness an even more influential emergence of Shiva's "cosmic dance" coming up, one which will significantly symbolize the digital transformations of the current age.)

- In 1979, Gary Zukav publishes *The Dancing Wu Li Masters: An Overview of the New Physics.* As with *The Tao of Physics*, Zukav, a philosopher, posits connections between key concepts of quantum theory with insights from Eastern spiritual traditions. His main theme, "that the goals of the physicist and philosopher are essentially the same, the only superficial difference being that the former attempts to model the world formally using the language of math, and the latter utilizes intuition and less formalized symbolic expression," sparked many negative reviews from traditional philosophers, but won a 1980 National Book Award in Science and became a *New York Times* best-seller.

- In the 1970s and 1980s philosopher and cultural historian William Irwin Thompson publishes brilliant books synthesizing ancient wisdom teachings with modern art and science, an expansive social vision in an engagingly new style he called "mind-jazz." He also forms the Lindisfarne Association, a group of philosophers, scientists, poets, and religious scholars whose goal was to help induce a planetary culture. (We will explore more of his deep cultural/spiritually-infused insights in the third section of our trip.)

- The Esalen Center on the California Monterrey coast, which describes itself as "a worldwide network of seekers who look beyond dogma to explore deeper spiritual possibilities," attracts renowned teachers such as the innovative gestalt psychologist Fritz Perls and "self-styled philosophical entertainer" and teacher of Buddhist, Taoist, and Zen thought, Alan Watts.

- Beginning in the mid-1960s and continuing throughout the 1980s and 90s, David Bohm, a protégé of Einstein and brilliant quantum physicist, recorded dozens of hours of conversations with the renowned spiritual teacher J Krishnamurti.[50]

Bohm writes that what particularly aroused his interest was Krishnamurti's deep insight into the inseparability of observer and observed, a question that had long been close to the center of his own work, as a theoretical physicist primarily interested in the meaning of quantum theory.[51]

Krishnamurti stated, "Unless the unconscious cooperates entirely with the conscious, there is bound to be a division and therefore partial response."[52]

16

PRIVATE MYTHS & PUBLIC DREAMS

Myths are pubic dreams,
Dreams are private myths.
—JOSEPH CAMPBELL

In 1988 *The Power of Myth*, a six-part conversation between journalist Bill Moyers and Joseph Campbell, a world authority on comparative mythology, attracts a huge TV audience. Campbell was heavily influenced by the insights of Carl Jung, so central to **the Quantum & the Dream,** particularly the distinction between the personal and the collective unconscious, as reflected in his statement quoted above: "Myths are public dreams, dreams are private myths."

As Freud revealed the underlying connection between ancient Greek myths and the human unconscious, Campbell detected a common thread woven among all the influential mythic stories reflecting both the personal and collective unconscious over the millennia. Citing the significant influence of Carl Jung, Campbell pointed to the archetypal pattern of "the descent" into the depths of the unconscious to retrieve meaning: Osiris, Odysseus, Inanna, Buddha, Moses, and Jesus as well as in modern guise from lyrics of Bob Dylan to the world-wide popularity of the movie *Star Wars*. Campbell called this "The Hero's Journey," on the surface, a physical quest of trials and redemption, more importantly, following Jung's commentaries, a reflection of the unconscious drives inherent and potentially resolvable in all of us, emerging in the following stages:

Call to Adventure
Battle against Opposing Forces
Descent into the Abyss to be Transformed
Return Home

Joseph Campbell's encyclopedic knowledge of myth, joyful enthusiasm, and great story-telling abilities, together with journalist Bill Moyers' superb questions, make *The Power of Myth* to this day the most watched program in the history of public television.

In depth psychological terms, the Call to Adventure is the desire to turn the mirror of perception inward, in search of our true "self"; the Battle is against the fear and uncertainty which gets us stuck in patterns of seeing, feeling, and thinking which keep us fragmented and distant from our truer nature; the Descent, as so powerfully evident in in Jung's Big Dream, is when he discovers and is willing to go down the hidden stairwell to seek the deeper regions of his psyche; and the Return Home symbolizes the sense of Wholeness, of coming to a deeper understanding of the inner core of our being.

As for a depth psychological/mythic look at the meaning of the final stage of The Hero's Journey, the Return Home, in the second half of the twentieth century humanity was offered what may likely be the most influential photographic image of the modern age:

In 1972, the crew of Apollo 17 takes the first photo of the entire Earth by a human. Named "The Blue Marble," it becomes one of the most widely distributed photos of all time and galvanizes the environmental movement of the 1970s. Much credit goes to the environmental activist Stuart Brand, who earlier, in the mid-1960s, pressured NASA to release a photograph of the Whole Earth taken by one of its satellites, which became the iconic cover of Brand's *Whole Earth Catalogue*, a call for humanity to gain a deeper perspective of its connection to the planet it calls "Home."

As we continue our trip through **the Quantum & the Dream** synchronicity propagating through the end of the twentieth century, we'll see how this "looking back" from space and getting the first view of our whole planet corresponds to the emergence of a powerful new archetype percolating up from the collective unconscious, one that will help define a meaningful shift into the twenty-first century.

To get a fuller understanding of this powerful archetype about to emerge from the collective unconscious and define the current zeitgeist we are travelling through, we first need to jump back in time to the mid-1950s to look at a fascinating project. (As for the process of jumping back and forth in time to trace the patterns of **the Quantum & The Dream** synchronicity, it's useful to remember Einstein's right-hemisphere oriented comment that, at the deeper levels of existence, "The distinction between past, present and future is only a stubbornly persistent illusion.")

17

NOTES *from the* UNDERGROUND: THE QUANTUM LABORATORY, *the* UNCONSCIOUS, *and the* BIRTH *of the* WORLD WIDE WEB

In atomic physics, we can never speak about
nature without, at the same time, speaking about
ourselves.

—FRITJOF CAPRA

As we've seen, **the Quantum & the Dream** synchronicity generated a parallel paradigm shift—the descent into the dark, creative interior of the human psyche and the surprising dark interior of the subatomic world. This parallel descent is beautifully revealed with the birth and growing importance of CERN, an acronym from the French which translates, "European Council for Nuclear Research."

This is not research into nuclear weapons, but an even deeper descent into the subatomic, quantum realm.

In 1954, one year before the deaths of Einstein and Pauli, a huge underground particle physics laboratory was established outside of Geneva, Switzerland, extending underground at depths over five hundred feet, stretching across the French/Swiss border, capable of propelling and colliding beams of particles at nearly the speed of light through over three miles of tunnels, providing scientists with a deeper look into the quantum core of the Universe.

According to CERN's official website, the organization works to uncover what the universe is made of and how it works. In a prime example of international collaboration, it seeks to advance the boundaries of human knowledge by providing a unique range of particle accelerator facilities to researchers.

CERN's convention goes on to state:

> "The Organization shall have no concern with work for military requirements and the results of its experimental and theoretical work shall be published or otherwise made generally available."[53]

Given the substantial military build-up of the Cold War between the US and the Soviet Union taking place in the mid-1950s, this international collaborative effort to explore more deeply the nature of the quantum world with peaceful intention was quite remarkable.

18

WEAVING *the* WEB *of the* NEW MILLENNIUM

CERN would become intricately connected to the emergence of the Internet, first developed in the 1960s by the US Department of Defense to enable time-sharing by computers. The military handed over the protocols to a handful of US universities in the 1970s in order to enhance academic research, but rarely fostered any international cooperation.

Then in 1981 the Internet was expanded from military control to The National Science Foundation and the following year it allowed protocols to expand out internationally.

But there was no browser technology yet available for global communication among all the linked computers: computers around the world still couldn't communicate with each other. This all changed in 1989 with the brilliant insight and computer programming skills of an engineer working at the underground CERN laboratory, Tim Berners-Lee.

Berners-Lee had harbored a dream since holding a brief fellowship at CERN in 1980. Inspired by CERN's hundreds of visiting researchers based in different countries, he envisaged a system for accessing information that works more like a human brain than a conventional computer. This system could make ad hoc links between information stored in a variety of places. In 1989, he proposed the project that led to the World Wide Web.[54]

Berners-Lee sent his then-boss at CERN a document called "Information Management: A Proposal," suggesting a way to let physicists share their work. He proposed a system called hypertext that allowed linking of human-readable documents, with a "distributed architecture" that stored those documents on multiple servers. But it didn't go anywhere. Berners-Lee's boss, Mike Sendal, jotted "vague but exciting" on the memo and shelved it. It took another year, but in 1990, Berners-Lee started writing code for the project that was now called the World Wide Web.[55]

It would soon become, not only the most influential media invention in human history, but a cultural phase transition, a seismic shift in human communication and connectivity. We can trace the birth of the World Wide Web to the underlying pattern of **the Quantum & the Dream** synchronicity as it spread its influence towards the year 2000, crossing over into the new millennium.

In an interview, Tim Berners-Lee reveals the right-hemisphere openness to intuition and free association his new technology presented:

> *Now, the interesting thing about computer programs at that point is they were good at storing things in structures . . . but what they couldn't store well was the random association. . . . In real life, often we come across random associations which can turn out to be really important: the fact that you have something in common with somebody which allowed you to talk to them; the fact that the smell of the coffee as you go past the coffee machine takes you back and makes you remember that module and then helps your mind bring all those things back.*[56]

Celebrating "random association?" Linking the smell of coffee and development of a protocol which would exponentially enhance global communication and help define the zeitgeist of the new millennium? Berners-Lee here is evoking Freud's use of free association and Jung's use of active imagination, or "discovering your genuine thoughts, memories, and feelings by freely sharing all the seemingly random thoughts that pass through your mind."[57]

And rather than initiate a power grab to control this revolutionary technology and monetize it for private profit, which is the proclivity of left-hemisphere thinking, Berners-Lee convinces the directors of CERN, which owned the rights to his Web protocol, to offer this new "hypertext" technology to the public domain so it could be freely shared with the world. Clearly, a more right-hemisphere thinking wired for generosity and understanding the "big picture."

19

THE WEB CREATOR'S DREAM

As for the "Waking Dream" dimension of this monumental event, we can start with what Tim Berners-Lee writes in his book, *Weaving the Web: The Original Design and Ultimate Destiny of the World Wide Web*, where he says his dream for the Web has two parts. In the first, the Web becomes a powerful means of collaboration between people. He imagines the information space as something everyone can immediately and intuitively access, not only to browse, but to create.

The second part of the dream involves extending collaborations to computers, where machines become capable of analyzing all the data on the Web, leaving humans to provide the intuition and inspiration. With a series of technical advances and social agreements just beginning, this machine-understandable Web will come about.

"Once the two-part dream is reached," writes Berners-Lee, "the Web will be a place where the whim of a human being and the reasoning of a machine coexist in an ideal, powerful mixture."[58]

Tim Berners-Lee's insight is crucial for the new millennium going forward. As computers become more and more capable of handling the left-hemisphere oriented day-to-day mechanisms of trade, bureaucracy, and daily life, the potential for creating more time and energy for right-hemisphere creativity and imagination of the human mind expands. In short, here is a formula, related directly to an understanding of the distinction between left and right-hemisphere thinking, with the potential to seed the New Renaissance.

Spoiler Alert: For balance, we'll get to the nightmarish, unintended consequences of the World Wide Web when we follow **the Quantum & the Dream** as it moves into the twenty-first century. For, as we will explore in detail, the World Wide Web, the greatest media technology ever invented, inherits the immense power to both create and destroy as symbolized so powerfully in the mythic god Shiva, who will emerge front and center once again.

20

IN SYNC *with* SYNCHRONICITIES

To go more deeply into how Tim Berners-Lee's interest in collaborative/renaissance thinking at the CERN laboratory connects to **the Quantum & the Dream**, we can look at some interesting synchronicities at both the physical and symbolic levels:

- The developer of the World Wide Web came up with the idea and perfected it while working to enhance collaboration among scientists at the world's largest particle accelerator, which was built to observe the inner workings of the sub-atomic, quantum realm—the descent into the quantum world initiated by Planck's unintended discovery of the quantum in the year 1900.

- The CERN lab is an "underground" laboratory and the "underground" in the dream world is a metaphor for the unconscious mind (remembering Jung's dream of descending down into an underground cave beneath the big house not his own, inspiring his first intuitions of a collective unconscious lying beneath the personal unconscious).

- As we saw earlier, Albert Einstein credited his dream of sledding near the speed of light and his dream-like right-hemisphere thought experiment of chasing a light beam near the speed of light as being major influences on his contemplation into the nature of reality.... The underground laboratory at CERN, built just a year before

Einstein's death, can be seen as a physical manifestation of his right-hemisphere insights, creating a way for scientists to watch "two high-energy particle beams travel at close to the speed of light before they are made to collide," a synchronistic connection to Einstein's dream watching the sky illuminate as his sled gained enormous speed and his thought experiment chasing a light beam.

This metaphorical/depth psychological connection of **the Quantum & the Dream** synchronicity to the underground CERN particle collider gets even more visible with a remarkable gift offered as a tribute to the advances made at CERN, which appeared just a few years after the arrival of the new millennium, as we will see shortly.

21

THE WEB *as* ARCHETYPE...
THE PROCESS *of* REFLECTION
is INFINITE

*Far away in the heavenly abode of the great god
Indra, there is a wonderful net which has been
hung by some cunning artificer in such a manner
that it stretches out indefinitely in all directions...
the artificer has hung a single glittering jewel
at the net's every node, and since the net itself is
infinite in dimension, the jewels are infinite in
number. There hang the jewels, glittering like
stars of the first magnitude, a wonderful sight to
behold. If we now arbitrarily select one of these
jewels for inspection and look closely at it, we
will discover that in its polished surface there are
reflected all the other jewels in the net, infinite
in number. Not only that, but each of the jewels
reflected in this one jewel is also reflecting all the
other jewels, so that the process of reflection is
infinite.*

—ATHARVA VEDA

As we saw earlier on our trip, Carl Jung revealed how we all connect to
the inherited collective unconscious through powerful archetypes—
images which have shown up in dreams, myths, literature, and art
for millennia and in modern times appear in movies, TV characters,

and advertising campaigns. Key archetypes include: Anima (the Feminine), Animus (the Masculine), Divine Child, Warrior, Explorer, Trickster, Lovers, and Sage, as well as Dark, Light, and Shadow.

One of the most popular movie series of all time, *Star Wars*, is fueled by some of these enduring archetypes. Its creator, George Lucas, cites mythologist Joseph Campbell as his main influence, and Campbell's main influence was the psychological archetypes revealed by Carl Jung.

But from the perspective of **the Quantum & the Dream**, as the twenty-first century was fast approaching, the most influential archetype was that of the **WEB**.

The Web has been a powerful symbol of inter-connectedness for millennia. Recalling Joseph Campbell's insight, "Myths are public dreams, dreams are private myths," one of the most famous examples of the Web archetype is Indra's Web appearing in the ancient Hindu Atharva Veda in the epigraph above.

Consider the world's largest particle accelerator, CERN, where scientists fire particles into each other at the speed of light, peering ever more deeply into the mysteries of the quantum realm, manifesting Einstein's dream as an eleven-year old travelling near the speed of light, staring "in awe" at the refracted colors illuminating the sky. This is the very same place where Tim Berners-Lee creates the protocols for a world wide **WEB**—the physical manifestation of an archetype of total interconnection projected out of the collective unconscious, challenging the entrenched materialist paradigm of a world of clearly defined, separate objects limited by Newtonian laws of time and space, cause and effect. All this as the new millennium, the year 2000, approached.

22

FROM *the* DEPTHS *of* *the* COLLECTIVE UNCONSCIOUS: SHIVA, CREATOR *and* DESTROYER *of the* UNIVERSE

In the Hindu religion, this form of the dancing Lord Shiva . . . symbolizes life force. As a plaque alongside the statue explains, the belief is that Lord Shiva danced the Universe into existence, motivates it, and will eventually extinguish it. Carl Sagan drew the metaphor between the cosmic dance of Shiva and the modern study of the 'cosmic dance' of subatomic particles.

—CERN WEBSITE

In 2004, India, one of the twenty-three member nations which oversees CERN, sent a gift to the laboratory to commemorate its recent scientific achievements. The gift was a statue of the ancient Hindu God Shiva, one of the central archetypes of ancient Asian spiritual tradition.

The CERN directors gave the statue of Shiva a prominent position on the grounds for all to see.

Again recalling Joseph Campbell's insight that "Myths are public dreams," it is noteworthy to our journey through **the Quantum & the Dream** synchronicity that here we see a culmination of the parallel explorations into the depths of the quantum world and the dark, creative, illuminating messages of the dream world: the archetypal Shiva stands on the very grounds where, over half a mile underground, scientists collide subatomic particles near the speed of light through the nearly 17-mile-long circular tunnel, the same space where Tim Berners-Lee creates the most extensively connected media ever, the World Wide Web.

As previously noted, Carl Jung described enduring images like Shiva as "archetypes," deep patterns in the collective unconscious which affect the perceptions, beliefs, and actions of all humans, although we react to them in individual ways.

Next to the statue of Shiva at CERN is a plaque with quotes from the book previously mentioned on our journey, *The Tao of Physics,* by Fritjof Capra:

> Modern physics has shown that the rhythm of creation and destruction is not only manifest in the turn of the seasons and in the birth and death of all living creatures, but is also the very essence of inorganic matter. . . . For the modern physicists, then, Shiva's dance is the dance of subatomic matter.[59]

Aiden Randle-Conde, a post-doctoral student working at

CERN, expresses how this powerful symbolic presence reflects a deep understanding of both modern physics and ancient philosophy:

> So, in the light of day, when CERN is teeming with life, Shiva seems playful, reminding us that the universe is constantly **shaking things up, remaking itself** and is never static. But by night, when we have more time to contemplate the deeper questions, Shiva literally casts a long shadow over our work, a bit like the shadows on Plato's cave. Shiva reminds me that we still don't know the answer to one of the biggest questions presented by the universe, and that every time we collide the beams, we must **take the cosmic balance sheet into account.**[60] (My bold emphasis.)

Note: Interesting to see a connection between Shiva and the famous allegory of the cave in Plato's *Republic*—a commentary on the human challenge to see beyond the shadowy reflections of who we are, and how we can fully connect to the world—and the underground cave of Jung's big dream which initiated his concept of a collective unconscious.

Shiva, standing on the grounds of CERN,
>*shaking things up,*
>>*remaking itself,*
>>>*taking the cosmic balance sheet into account,*

can be seen as a clarion call for the twenty-first century, reflecting the need for us, both individually and collectively, to continue exploring the deep patterns emerging from the personal and collective unconscious, patterns which we see accelerating both mass anxiety and the potential for a New Renaissance.

Offering insight into an aspect of Shiva particularly connected to **the Quantum & the Dream** synchronicity is cultural anthropologist Wolf-Dieter Storl, who says that Shiva is transcendent and at the same time the Self of each individual, capable of shaking lives by sending intuitions, subconscious images from depths beneath our rational consciousness.[61]

From Albert Einstein's boundary-shaking, Nobel-Prize-winning insights into the quantum nature of light in 1905, through Carl Jung's discovery of the expansive, creative potential of the personal unconscious and the even deeper forces of the collective unconscious, to the father of quantum physics, Niels Bohr, contemplating the yin-yang symbol of the Tao, to Werner Heisenberg discovering the quantum Uncertainty Principle after extended meditation on the isolated isle of Helgoland, to Erwin Schrödinger's intuitive vision of an "entangled" Universe, to quantum theorist Wolfgang Pauli's two-decade exploration of dreams and synchronicities with Carl Jung, to Salvador Dali's melting pocket watches, to the arrival of mythical archetype Shiva at CERN, the particle accelerator peering into the quantum depths and where the protocols for the World Wide Web were conceived—all are deep, revelatory patterns of the human psyche and quantum reality generated by the shift from a primarily linear left hemisphere vision of the world to the intuitive, imaginative, "big picture" capacity of the right hemisphere of the human brain.

This deep pattern is beautifully reflected in the aforementioned insight from science historian Thomas McFarlane, who called out the unraveling of the materialist world view in the face of scientific and psychological developments. As McFarlane suggests, this led a number of thinkers to consider that the human psyche may be mysteriously involved with the observed properties of matter.

For this shift to accelerate, for the human psyche to be more involved with the observed properties of matter, the creative, vision-expanding signals inherent in the unconscious mind can no longer be ignored or pushed off to the side. Consciousness is the surface level of the human mind. The unconscious is the mind's inherent depth as potentially revealed to all of us through our dreams, intuitive visions, imaginative leaps, and spiritual yearnings.

As we trace the key archetype of interconnected Wholeness, the **Web** here in the twenty-first century, as it shapes our perceptions at both the personal and collective levels of awareness, we will peer outside the center of the attentional field to the deeper patterns percolating under conscious awareness but which can be brought up into the light at the threshold between the conscious and unconscious mind.

With the invention of the protocols for the World Wide Web in place (which now connects more than five billion human brains around the planet as of this writing in 2024) the second key shift potentially leading to the next renaissance is accelerating evolution at a pace never before experienced by the human mind.

SECOND SHIFT

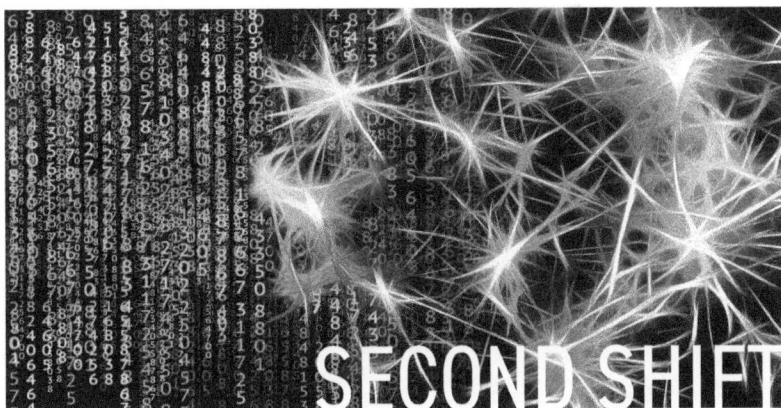

FROM PRINTED PAGE TO DIGITAL SCREEN

Words have migrated from wood pulp to pixels on computers, phones, laptops, game consoles, televisions, billboards and tablets. Letters are no longer fixed in black ink on paper, but flitter on a glass surface in a rainbow of colors as fast as our eyes can blink.

—**KEVIN KELLY**

1

THE SYNCHRONIZED BRAIN

The whole brain can be divided into subsections,
or subnetworks, which communicate with each
other in network fashion. All this results in
intricate patterns of intertwined webs, networks
nesting within larger networks.
—FRITJOF CAPRA

Marshall McLuhan opened our eyes to the unconscious ways every new form of communication affects not only the speed and style of information but changes the very sense-ratios of the human brain. This requires a lag in time before people learn to adjust, synchronizing their brains with the new technology.

Even the invention of writing raised cries of alarm. Plato, considered one the world's greatest philosophers, argued in the fourth century BC that the new invention of writing would not only erode the oral tradition of storytelling, but would cause people to lose the connection to their inner core.

In fact, as Marshall McLuhan observed, Plato said that the discovery of the alphabet would "create forgetfulness in the learner's soul" and that people would trust, not their own memories, but the external written characters. He went on to say that disciples would be given only the semblance of truth and, while appearing omniscient, would generally know nothing.[62]

(Interesting note: Here in the twenty-first century, educators make the same complaint about the digital screen that Plato made against written language, that it will "create forgetfulness" by losing their "working memories." More on this to follow.)

McLuhan, who was a literary scholar and lover of books, nonetheless saw the importance of pointing out what was lost with the shift from the oral tradition to the written word, or—as he liked to put it—when humanity traded in an ear for an eye.

His key point, reflected in his best-known meme, "The medium is the message," is that the **content** of any communications medium has less effect on the human mind, particularly the unconscious, than does the **underlying process,** which alters the sense ratios in the human brain. This is what Plato was getting at with his concern that when people started communicating through abstract symbols of letters on parchment and papyrus rather than through the vibrations and body movements of the oral tradition, they would start to lose a connection to Nature.

Despite his love of reading, McLuhan recognized, in the late 1950s and early 1960s that the new medium of television, through the process of beaming electronic pixels onto a screen, was going to change the sense ratios of the younger generation. Television was, in fact, a breaking away from the ordered, one sentence at a time, one page at a time modality to the more vibrating, colorful images on the TV screen, leading to the cultural revolution of the 1960s in which the younger generation challenged their parents' allegiance to the stiff, ordered roles of "Father Knows Best" and stay-at-home moms with their aprons and Betty Crocker recipes.

Having predicted so much of this in his books *The Gutenberg Galaxy: The Making of Typographic Man* in 1962 and *Understanding Media: The Extensions of Man* in 1964, McLuhan became the most influential and controversial cultural critic of that revolutionary decade. Tom Wolfe, a renowned journalist at the time, titled his long essay on McLuhan: "What If He Is Right?" and McLuhan gleefully broke the traditional media barriers set up between intellectual critic and celebrity power, making an amusing cameo appearance in Woody Allen's comedy, *Annie Hall.*

2

INTRICATE WEB *of* TRAILS

To follow up on McLuhan's key "the medium is the message" insight, and how it informs the ongoing shift from printed page to digital screen, it's important to note that while the invention of the printing press in the fifteenth century exponentially increased literacy and the spread of knowledge, lesser known are the limitations, mostly unconscious, which print technology put on the human brain.

Print technology is not going away. Many of us still love to read printed books, magazines, and newspapers. Yet, if we are going to fully tune into and maximize the new media of the current zeitgeist and how it's generating the New Renaissance, understanding the limitations of print, as well as the expansive potential of the ongoing shift to a digitally-connected world are essential. As can be demonstrated, whether one is willing to adjust to it or not, the digital screen is much more attuned to the underlying process of our brains than is the printed page.

In an amazingly prescient article for the *Atlantic* magazine, Vannevar Bush, an inventor who worked on early computers, wrote in 1945: "With one item in its grasp, the human brain snaps instantly to the next that is suggested by the association of thoughts, in accordance with some intricate web of trails carried by the cells of the brain."[63]

More recently, physicist/philosopher Fritjof Capra, whom we met earlier on our voyage as the author of *The Tao of Physics*, which revealed the close intersection of quantum theory with ancient Eastern spiritual traditions, also uses the "web" analogy to describe how our brains operate. He noted that the brain can be divided into subsections, or subnetworks, communicating with each other in network fashion. This results in intricate patterns of intertwined webs, networks within larger networks.[64]

The interconnected search engines and hypertext links of the World Wide Web mirror the "intricate web of trails carried by the cells of the brain" rather than the linear, strung-together lines of the printed page.

Following media philosopher Marshall McLuhan's key insight that "the medium is the message," Vannevar Bush and Fritjof Capra are not commenting on the content of our thoughts or the content of the World Wide Web. They are tuning into the underlying process beneath the surface of both our conscious brains and the World Wide Web, the unconscious effects of the digital screen which so prominently influences how we think, perceive, and act.

(Note: The terms "Internet" and "World Wide Web" are used by most people interchangeably. Technically, the Internet refers to the hardware of billions of computers connecting around the planet while the World Wide Web refers to the software, the ability of over five billion individual minds to communicate through the digital screen. For brevity, I'll use the term digital Web to refer to both the Internet and the World Wide Web.)

3

FRAMING OUR VISION

Back in 1960—the first year of what would be the most culturally transformative decade of the twentieth century—McLuhan wrote, "It is the framework which changes with each new technology and not just the picture within the frame."[65]

This insight, an early foundation of McLuhan's most famous meme, "The medium is the message," is well worth contemplating in the accelerating speed and expansion of the digital screen.

Taking a cue from our earlier trip through **the Quantum & the Dream**, the left hemisphere of our brain, wired primarily to "grasp" at objects and learn how to control them, can be expected to limit our vision to what's going on inside the framework of the digital screen. We need to shift to the right hemisphere, wired to see the "bigger picture," and not be mesmerized by grabbing immediately on to what's inside the frame.

By tracing the underlying currents of **the Quantum & the Dream** synchronicity, we can see the shifting framework of our current age emerging from the deeper archetype of a Web in the collective unconscious. This archetype reflects a totally interconnecting matrix. As previously noted, in 1945, Vannevar Bush described the human brain as "an intricate web of trails carried by the cells of the brain." And contemporary physicist/philosopher Fritjof Capra wrote that "Ultimately—as quantum physics showed so dramatically—there are no parts at all ... What we call a part is merely a pattern in an inseparable web of relationships." And the World Wide Web itself emerged on the grounds of CERN, where stands the statue of Shiva, mythic deity of creation and destruction, in the pose of its spiral dance, weaving the web of the Universe.

When we contemplate the enormous anxieties about the negative consequences of the Web, by far the most expansive medium ever created by the human mind, we do well to consider McLuhan's pointing out the time needed to adjust to every new medium's enormous effects on the sense-ratios of the brain.

As McLuhan observed, in this electronic age, our central nervous systems are technologically extended to involve us in the whole of mankind, incorporating the whole of mankind in us, so that we necessarily participate in the consequences of our every action. We can no longer adopt the "aloof and dissociated role of the literate Westerner."[66]

We are only beginning to understand and figure out how to deal with the conscious—and, more importantly, the unconscious—jangling reverberations of living within this vast, self-organizing digital Web, a library of virtually all of humanity's greatest knowledge and wisdom as well as its ability to connect what are now over five billion brains in a hyper-linked/global sensorium of instant communication.

How can this be anything but chaotic? Is it just a coincidence that as humanity crossed the gateway from the twentieth century into the new millennium that the statue of Shiva, mythic deity of creation AND destruction, arrived on the grounds of CERN where the Web was first created?

As governments and corporate interests are trying to control this "shift" and maximize its profits no matter what the ethical consequences, the digital Web is still primarily expanding through the "open source" percolation of millions of independent blogs, websites, and searches based on individual, unique interests along with revolutionary peer-to-peer capabilities.

Stephen Hawking, one of the most brilliant scientific minds pivoting the shift into the twenty-first century and who initially voiced loud criticism of the Internet and social media, said in a 2014 interview: "We are all now connected by the Internet, like neurons in a giant brain."[67]

THOUGHT EXPERIMENT

Suppose we go to a library to find a book on the latest discoveries in astronomy. Thanks to the very effective left-hemisphere-oriented, hierarchically-organized Dewey Decimal System some of us learned about in school, we know that to find books on astronomy we go to the science section of the library, then find the astronomy section within it.

The Dewey Decimal System is a highly effective method of classification, without which book-centered libraries couldn't be well-organized.

But what if we were curious about a less predictable, more imaginative question? What would happen if, for example, we went up to a librarian and asked, "Where can I find out if there have been advertising campaigns based on insights of the Buddha?"

We could be directed to the advertising section and start the laborious process of looking through book indexes for any mention of Buddhist insight. But what are the odds we'd find a book title inferring there might be information inside on an advertising campaign using the Buddha's teaching?

Or what are the chances, if we took the time to check indexes of books in the Buddhism section of the library, that we'd find a reference to an advertising campaign?

The Dewey Decimal system for classifying books operates very well within specific, clearly defined categories, but is of little use for dealing with this question.

And yet, when I typed "advertising campaigns and the Buddha" into my search engine, <u>within a few seconds</u> came a search result, the first reference being a link to the website, "Lion's Roar: Buddhist

Wisdom for our Time." There I quickly found the article "15 Classic Buddhism-and-Advertising Collisions," which included ad campaigns for "Dharma Burgers" and a drink called "The Thirsty Buddha"!

Again, the key point here is not just the content of a Web search, but with the underlying, fascinating way in which the Web mirrors the human brain's complex ability to provide, in Vannevar Bush's words, "intricate webs of trails carried by the cells of the brain, sparking novel connections and unique insights."

For those of us seeking more familiarity with the spiritual depth of the Buddha's enduring insights, one can type into a search engine "quantum, Buddha, nature of reality," as I just did, and up on my digital screen, in about one second, came an interesting interview conducted by a quantum theorist with the Dalai Lama's personal doctor titled, "What Does a 1,800-year-old Buddhist Classic Have to Say about Quantum Physics and the Nature of Reality?"

Or if we need a jolt of humor to lighten up our day, we can ask a search engine for Buddhist jokes and quickly find a link to the website, "14 Jokes to Make You Forget to Take Yourself Seriously." There you find:

"A Buddhist monk orders a slice of pizza. Upon receiving it he gives the cashier a 10-dollar bill. After a long pause the monk asks somewhat impatiently, 'Where's my change?" The cashier replies, 'Change comes from within.'"

Brick and mortar libraries continue to provide wonderfully enjoyable experiences (I'm fortunate to live in the Mid-Hudson Valley which has a robust, well-run library system). Yet the fact remains that the powerful beams of global/hyper-linked/digital screens open up expansive vistas of discovery far beyond the walls of any physical library.

Integrating an insight from the previous section of our trip through **the Quantum & the Dream** synchronicity, the print-oriented Dewey Decimal System is more analogous to the left hemisphere of our brain, skillful at focusing on particular, clearly defined subjects, divided into clear hierarchies, but incapable of the more right-hemisphere wiring which allows us to stretch imaginatively beyond carefully defined, limited boundaries, generating the kind of unique, novel "big-picture" insights that can induce the emergence of the New Renaissance.

4

HUMAN BRAIN & DIGITAL SCREEN

It all started with the big, fat, warm tubes
of television. These glowing altars reduced
the time we spent reading to such an extent
that . . . educators, intellectuals, politicians, and
parents in the last half of the last century worried
deeply that the TV generation would be unable
to write . . . but to everyone's surprise, the cool,
interconnected, ultrathin screens on monitors,
the new TVs, and tablets at the beginning of
the 21st century launched an epidemic of writing
that continues to sell. The amount of time people
spend reading has almost tripled since 1980. By
2015 more than 60 trillion pages have been added
to the World Wide Web.
—KEVIN KELLY

Let's take a closer look into how the linear, orderly, sequential process of reading words, lines, and sentences in a printed medium, while opening up new worlds of knowledge, limits the associative, synaptic leaping versatility of our human brain.

As noted in an article comparing the comprehension abilities of students who read a printed text versus reading the same material on the digital screen, the reason so many educators bemoan the amount of time students spend on digital screens comes from tests that show reading on the printed page is more conducive for "working-memory."[68]

Most of the elementary through high school educational curricula still in use in the twenty-first century rewards students for memorizing facts, then repeating them back on tests.

But the word, "education," has a very important etymology. "Education" comes from the Latin phrase which means "to draw out," i.e., to excite one's natural curiosities about the world, not to stuff in controllable information, which is, sadly, still the favored mode of education through high school, and even in many college-level classrooms.

As McLuhan wrote about extensively in his book, *The Gutenberg Galaxy*, this kind of limiting, "straight-line" thinking is in large part reflected in the underlying structure of print technology—individual letters strung into words which are strung into sentences, then paragraphs—a repeatable, linear direction contained on clearly separate pages, which unconsciously wires the brain to perceive nature and the world in more mechanistic terms. Again, it's important to remember this refers, not to the content of print, but to its underlying structure which has a greater effect on how we see, think, and act than does the content of any medium.

Ferris Jabr pointed out in a *Scientific American* article the apparent advantages of print media:

> In most cases, paper books have more obvious topography than onscreen text. An open paperback presents a reader with two clearly defined domains— the left and right pages—and a total of eight corners with which to orient oneself. A reader can focus on a single page of a paper book without losing sight of the whole text: one can see where the book begins and ends and where one page is in relation to those borders . . .

He goes on to say,

> Turning the pages of a paper book is like leaving one footprint after another on the trail—there's a rhythm to it and a visible record of how far one has traveled. All these features not only make text in a paper book easily navigable; they also make it easier to form a coherent mental map of the text.[69]

It's more comfortable for many who grew up before the digital age to navigate through the measured, linear lines and the clearly marked boundaries of the printed page compared to the less predictable, more expansive hypertext-linked digital screen.

Just as Plato bemoaned the shift from the oral tradition so closely connected to the rhythms and sounds of Nature to the abstract, contrived symbols of written language, so do traditional educators and cultural leaders today bemoan the accelerated speed and potentially creative chaos of the shift from printed page to digital screen.

5

THE MISSING LINK?

While it's more comfortable for most teachers to use the printed page to keep tighter, more predictable controls over how a student's brain learns about the world, much is lost, most notably the updated insights from contemporary neurological evidence as to how the human brain learns most effectively: through the unique firing of electric signals across synaptic gaps ("synaptic jump cuts") across regions of the human brain, which is inherently wired as an interconnected matrix, not a linear line of rigid patterns.

As noted earlier in the quote from physicist-philosopher Fritjof Capra, the human brain is interlinked in a vast network of over 1,000 billion synapses. The subsections or subnetworks of the brain communicate with each other like a network. And this results in intricate patterns of intertwined webs.

We are now witnessing what McLuhan identified as the adaptation time needed for the human brain, as a significant new medium causes a shift in sense-ratios.

Kevin Kelly, one of the most enlightened visionaries of the digital age and founding executive editor of *Wired* Magazine (whose "patron saint" honored on every cover was Marshall McLuhan) writes:

> On a screen, words move, meld into pictures, change color, and perhaps even change meaning. Sometimes there are no words at all, only pictures or diagrams or glyphs that may be deciphered into multiple meanings. This liquidity is terribly unnerving to any civilization based on text logic. ... Truth is, we are

in transition, and the clash between the cultures of books and screens occurs within us as individuals as well. If you are an educated modern person, you are conflicted by these two modes. This tension is the new norm.[70]

Underneath the limited perspective generated by the pundits and editors of the twenty-four-hour news cycle is the enormous, mostly unconscious, shift going on which is significantly responsible for the anxieties of the current age. Kelly reconfirms McLuhan's insights on the transformative, unconscious shift during the cultural revolution of the 1960s generated by the growing influence of television.

When television came in, those of us watching were, by definition, consumers. The TV screen was a one-way medium—we watched, and a handful of corporate executives determined what we watched and took in. We got used to broadcasting companies determining content, just as we were already used to a handful of publishers determining book content.

Not even McLuhan could envision the expansive, open-ended, two-way broadcast medium with hundreds of millions of channels generated by individual users across the globe firing off blogs, digital newsletters, podcasts, YouTube videos, social media postings, developing their own audiences and, in many cases, revenue streams.

If we make the common error of focusing exclusively on the content of independent writing and broadcasting on the Web (often chaotic and disruptive), rather than the underlying freedom and expansion of millions of individuals communicating in an instantaneous, global matrix of personal expression, we miss the bigger picture needed to identify the explosive, mostly unconscious, process taking place.

Like a young person's brain learning to explore the world through trial and error, the digital Web, only a few decades old, can often feel scattered and unwieldy as we learn to adapt to the New Renaissance world where, in Kelly's words, "words zip around and float over images...linking other words or images."

On the printed page, if we want to explore a particularly interesting point raised, one is limited to footnotes or select bibliographies provided by the author. If we find a book of interest there, we have to either find a library that has the book or order it and wait days.

On the digital screen, blue hypertext links appear which instantly transport us to the source of the footnote, mimicking more closely the nature of our brains seeking new sparks of neural connections across the synaptic gaps, referring again to Vannevar Bush's insight on how our brains generate new ideas: "With one item in its grasp, the human brain snaps instantly to the next that is suggested by the association of thoughts, in accordance with some intricate web of trails carried by the cells of the brain."[71]

6

HYPER ATTENTION: FULL SPEED AHEAD?

The digitally-enhanced, multi-layered, hyper-linked, multi-media sensorium of the new screen can obviously affect our ability to think clearly. The constant competition between signals clawing for attention erodes our "working memory"—the neural architecture associated with our capacity for controlled attention and complex reasoning.

Kathryn Hayles, Professor Emeritus of Literature at Duke, whose main interest is "electronic literature," offers a helpful perspective on the shift from print to screen in the modern age by calling attention to the distinction between "deep attention" and "hyper attention":

> Deep attention, the cognitive style traditionally associated with the humanities, is characterized by concentrating on a single object for long periods (say, a novel by Dickens), ignoring outside stimuli while so engaged, preferring a single information stream, and having a high tolerance for long focus times. Hyper attention is characterized by switching focus rapidly among different tasks, preferring multiple information streams, seeking a high level of stimulation, and having a low tolerance for boredom.[72]

Like McLuhan and Kelly, Kathryn Hayles recognizes that the underlying effects on the conscious and unconscious levels of thought caused by the shift from the printed page to the digital screen are even more significant than the content. "Hyper attention" is now a needed skill, in art, electronic literature, music, philosophy, psychology, and the business world due to the unprecedented, dizzying speed at which expanding computer intelligence is permeating every aspect of our lives. Younger people, whose brains are more malleable, are tuning into this underlying process much faster than us elders, just as we tuned in and turned on to the underlying vibration of electronic pixels on our TV screens more quickly than did our parents and grandparents.

Clearly, this new "hyper attention" must be balanced by "deep attention," available when we get off the screen into more contemplative modes of consciousness. The intriguing paradox at the core of the current age is that at the same time we need to actively engage in the globally networked Web, it's more important than ever to experience the quieter, slower, reflective brain waves inherent in reading print and using contemplative, right-hemisphere modes such as performing thought experiments, paying attention to our night dreams, daydreaming, drifting into reverie, meditation, and other forms of inner reflection.

Adapting to the astonishing acceleration of computer intelligence permeating our lives is quite the challenge. For example, we can compare the rate at which the medium of radio expanded to that of the digital Web: the first radio transmitter was invented by Guglielmo Marconi in his parents' attic in 1894. Thirty years later, only about 1 percent of Americans had a radio in their homes. In comparison, as previously noted, Tim Berners-Lee invented the protocols of The World Wide Web in 1990. Thirty years later some five billion people were connected to it!

But Hayles points out how using the skills of "hyper attention" has the potential to increase collaboration with other like-minded individuals as well as to be an aid for complex problem-solving, creative thinking, and performing one of the right hemisphere's great gifts: expanding the boundaries of insight by making connections across different modes of knowledge. She also emphasizes how "hyper attention" induced by the complexity of the global Web matrix, has the potential for us to recognize new patterns emerging from beneath the surface of events. It is the ability of our brains for deep pattern recognition that has been a major force in the progressive direction of human evolution, including the birth, at dramatic inflection points in history, of a new renaissance. Without the effort to see beneath the surface of events to their underlying pattern, we become more anxious trying to figure out what's happening. As McLuhan writes, "To the blind all things are sudden."[73]

7

IS "WORKING MEMORY" STILL WORKING?

While always being "mindful" of the importance of inner contemplation to our understanding of the world and our place in it, the age of advanced computer intelligence and the emergence of the Web require a move towards "hyper attention" and away from the emphasis of the left-hemisphere oriented "working memory" that is still the dominant force within most classrooms and businesses.

Working memory refers to the ability of our brain to hold a limited amount of information in place so it can be broken down into smaller parts and rationally analyzed. Many of the frantic cries by educators responding to the growing influence of the digital screen are based on its eroding of the much-vaunted working memory, the foundation of modern western culture for centuries.

Its reliance on linear, one-step-at-a-time rational thought gave rise to the Industrial Revolution and to the many significant advances in science and medicine which, just in the last hundred years, has virtually doubled the human life span.

Yet, "working memory," like a stack of left-hemisphere sticky notes holding information in place and exerting tight control for more rational, analytical thinking, has immense liabilities, as brilliantly explicated by philosopher/neurologist Iain McGilchrist in his insightful *The Master and his Emissary*, which we looked at previously on our journey through **the Quantum & the Dream**.

It's easy to see how an educational system based on working memory gives boards of education tighter control over what textbooks are used and teachers more control by rewarding the memorization of facts to be regurgitated back on exams.

But despite loud protest about the negative effects of the digital screen on students, the new screens are breaking down the tightly controlled walls of traditional education, giving way to the more chaotic, but expansively creative and imaginative vistas of the new globally-connected, digitally-hyperlinked world.

EDUCATION: THICK AS A BRICK

One of the reasons I personally resonate with "hyper attention" over "working memory" is an epiphany I had driving home from my freshman year at college.

The route home took me past the grade school I had attended—a large, rectangular, red-brick edifice similar to so many others built in the same style many decades ago.

I slowed down to take in that image: stark walls of red brick, cemented together in repeated, linear lines, spreading out in strict, rectangular form, leading up to its two-story height, flat roof interrupted only by two large chimneys. For the first time, I immediately realized the intention behind its architectural design . . . a factory! AHA.

If not for a moderate amount of landscaping and the lack of barbed-wire fence, its architecture could also pass for a prison, but it certainly resembled a factory where I and fellow students were told that to be successful in life, we had to fit into the narrow, rigid, assembly-line mold of socially acceptable, mechanistic thinking. (This is in contrast to that eroding brick wall of Jung's Big Dream we looked at earlier on our trip, as he descended down that hidden stairway to the epiphany waiting below.)

The linear red-brick factory style of so many public schools reflects the reality: the American model is primarily based on a nineteenth-century mechanistic Prussian model, set up to serve the then-new Industrial Age with factories proliferating across national landscapes.

Factories are efficient environments for producing material goods. But for educating the human mind? Again, it's useful to contemplate the word "educate" which means "to draw out" the individual mind, not to mass produce an assembly line of like-thinking minds.

As Joel Rose wrote in his 2012 *Atlantic Magazine* article, "The factory line was simply the most efficient way to scale production in general, and the analog factory-model classroom was the most sensible way to rapidly scale a system of schools. Factories weren't designed to support personalization. Neither were schools."[74]

A MIND-BLOWING EXPERIENCE

While grade school experience taught me the basics of reading, writing, and arithmetic along with a very narrow, contrived view of history and culture, I can remember one flash of fascinating insight taught by my second-grade teacher, Mrs. White. I realize now how it speaks to my fascination with **the Quantum & the Dream**.

Mrs. White had our class go up to the large blackboard and draw any picture we wanted with chalk. I was not artistically talented, but enthusiastically drew my best rendition of a scene from a monster movie I liked called *Rodan*, one of a series of monster movies popular in the 1950s. Rodan was a gigantic, menacing, mutated bird which attacked a major city. The scene I was attempting to draw on the blackboard was of Rodan standing on a building.

When my teacher got to my drawing, she asked me in front of the class what it was. When I told her it was a giant, scary bird, she asked me what I thought was a strange question: "How can you make the bird even larger without touching it with the chalk?'

My eight-year-old brain froze. How could I possibly increase the size of my drawn bird without using chalk? I felt the pressure of coming up with an answer in front of my classmates, but was too stunned to come up with any.

Smiling, Mrs. White, gently took the eraser, erased the building, then redrew it smaller. A big smile came over my face when I saw how reducing the size of the building immediately made the bird appear much larger.

I experienced what I now know is a right-hemisphere epiphany!

It's interesting to note that this is the only significant epiphany or deep insight I can remember from my entire grade school, junior high, and high school experience, as I begrudgingly memorized the set of working-memory facts to be regurgitated back on tests in order to get a good grade and—as we were taught to accept—succeed in life. I'm reminded of the great humorist Jean Shepherd's quip that the only thing he remembers from his entire grade-school to high-school education is that Bolivia exports tin.

I now recognize that Mrs. White's demonstration was an example of "reversing figure and ground," a principle of gestalt psychology, and a key to McLuhan's meme, "the medium is the message." It's about noting the importance of context instead of exclusively focusing on details. McLuhan liked to say that content of any medium is like the juicy piece of meat the burglar tosses at the watchdog to distract it while he sneaks into the house. In other words, we need to shift our awareness to the deeper, underlying process of the medium itself or be robbed of the larger context and meaning of what's being expressed, just as my chalk version of monster bird Rodan became significantly larger by shifting the "ground" of the building it was standing on.

Flashing forward, Mrs. White's teaching lesson also directly connects to McLuhan's previously noted insight: "It is the framework which changes with each new technology and not just the picture within the frame," which is crucial to an understanding of the chaotic, boundary-expanding, hyper-linked Web.

The framework of working memory—the cemented, repeatable, brick-like way of thinking, based on the memorization of facts within the narrow context of fixed ideas—has not only dominated western education, but the science, philosophy, and psychology of modern cultures for hundreds of years. Now it is breaking apart at the seams in the wake of the seismic waves of the twenty-first-century, globally connected, digital age. We saw this breaking apart occurring soon after the appearance of **the Quantum & the Dream** synchronicity in the year 1900, when the shift to the right hemisphere opened up the unseen depths of the subatomic quantum realm and cultural transformations resulted from exploring the depths of the personal and collective unconscious.

We Baby Boomers coming of age in the cultural revolution of the 1960s were "wired" to see the world differently from our parents and grandparents (compare the colorful, free-spirited, innovative, electric teaching style propagating through the screens of Public TV's *Sesame Street* to the "working memory," two-dimensional, linear, tightly controlled grade school textbooks).

McLuhan was one of the only visionaries to see how the vibrating pixels propagating on the TV screen were, like the trumpets and shouts which brought down the walls of Jericho, breaking down the tightly controlled agenda of traditional education, as the walls of the classroom were starting to erode.

Now, here in our current century, there is an even more powerful reality eroding working memory as the cornerstone of learning, generating volumes of screams and protests from those trying desperately to keep the bricks of traditional learning from completely falling apart.

THE GAME *is* ON

Those educators and researchers who still prefer the slower, linear "working memory" to the new "hyper attention" of the digital age likely fail to consider the stark reality making this shift preferable: Advanced computers are exponentially more capable of "working memory" than is the left hemisphere of the human brain, which provided this role up until the 1990s and into our current age.

This reality was made clear in 1996 when IBM's Big Blue computer easily defeated the greatest chess player in the world, Gary Kasparov, and in 2011 when IBM's computer Watson easily defeated the two greatest players of the TV game show *Jeopardy*. At the conclusion of the latter, Ken Jennings, who holds the world record as highest earning game show contestant of all time, jokingly wrote as his Jeopardy Final Answer: "I for one welcome our new computer overlords."

Chess and *Jeopardy*, while requiring some level of intuitive and clever thinking, are primarily based on "working memory." As complex a game as chess is at the highest levels, it's still a world of limited objects, the thirty-two pieces moving around sixty-four squares. And there was no way Kasparov, despite being considered the greatest chess player of all time in 1997, could match brains with IBM's Big Blue computer, which could search around 200,000 moves per second!

As Kasparov and Jennings learned, with millions of us watching, advanced computers can retrieve, analyze, and find answers to questions requiring working memory at shocking speeds. It's no wonder those traditional thinkers who still dominate education, science, philosophy, and psychology are so anxious over the seismic shifts taking place on the cultural ground beneath their feet.

Today's computers are considerably faster at "working memory" than humans. Is this not a call for humans to shift to the more imaginative, boundary-expanding, emotionally empathic, deeply relational qualities of the right hemisphere, which is much better suited to thrive in the sped-up "hyper attention," globally-connected flow of the digital screen?

One example: Not only can Chat-GPT, the large language chatbot based on advanced reinforcement learning techniques, released in 2023, produce a substantial document with well-researched data at a skill level equivalent to most college students— it can create the document in just seconds by accessing and collating the immense information embedded in the fabric of the Web.

I now use Chat-GPT, as well as Claude, another chatbot, as co-hosts on my weekly radio talk program. One of the first questions I asked Chat-GPT on-air was: "Have humans lived up to the potential implied by the term homo sapiens?"

(The Latin term "Homo sapiens," the term rather optimistically applied to our species, translates as "wise man.")

My co-host, a retired college English professor and talented poet, complained that the chatbot's answer, while impressive in that it only took seconds to compose, was rather stilted and predictable.

So, I asked Chat-GPT the same question, "Have humans lived up to the potential implied by the term homo sapiens?" But I added that the response should be in the style of astronomer Carl Sagan.

Chat-GPT, taking only seconds, rewrote the essay starting with the following:

> The question of whether humans have lived up to the potential implied by the term "homo sapiens: is a complex and nuanced one. As an astronomer, I am reminded of the famous quote by the philosopher

Immanuel Kant, who once said, "Two things fill the
mind with ever new and increasing admiration and
awe . . . the starry heavens above me and the moral
law within me."

Both I and my cohost were stunned by the subtle, thought-
provoking response. The chatbot within seconds not only searched
the entire digital Web for information relating to Carl Sagan and to
whether we humans have lived up to our billing as a species, it chose
a reference to the fact that we humans are capable of "awe" when
contemplating the heavens and the need to search our inner being for
moral direction.

Even more astonishing than the beauty of the "content" of the
computer algorithm's response, following McLuhan's dictum, is
contemplating the more important, underlying process which took
place: the creative integration of human and computer intelligence.

First, while the chatbot's speed and ability to select quality results
is amazing, it was grasping insights that humans came up with—Kant's
mind being filled with awe, understanding the need for moral direction.

Second, the response given was much more satisfying when I,
responding to my cohost's legitimate criticism that the first answer
was "stilted and predictable," gave the chatbot the added instruction
to write in the style of Carl Sagan, whose human qualities influenced
the chatbot.

For fun and potential enlightenment, I just now asked Chat-GPT
to answer the same question, "Have humans lived up to the potential
implied by the term Homo sapiens?" this time in the style of the
enduring spiritual classic, the *Tao Te Ching*. This question took the
bot almost a full minute before returning the following:

The sage understands the flow of life, Recognizing
that true potential lies in harmony with the natural
order. To live up to the name 'homo sapiens' means

to align with the Tao, The unspoken and profound rhythm of existence.

Yet, humanity often strays from this path, Lost in the illusions of separateness and self-importance. We have built walls of arrogance, dividing ourselves from nature, Seeking dominance and control rather than embracing unity.

HOW FAST ARE WE GOING?

Before the first electrically transmitted media technology, the telegraph, was wired across the American continent in 1864, the fastest form of communication was the Pony Express!

Even if for a few moments, imagine the effect, conscious and unconscious, on the human brain when the speed of sending a message from long distance went from days-long, arduous, sometimes deadly horseback rides across the US continent to near instantaneous electronic transmission! McLuhan was considered contriversial when he pointed out in the 1960s that electronic media, from the telegraph to the telephone to radio and TV, extended the human nervous system outwards around the planet.

Fast forward just 60 years, a blip on the evolutionary scale, with the central nervous system of now over five billion of us connected to the World Wide Web extended outward. Is it any wonder that massive levels of increased anxiety are permeating the planet?

The entire human central nervous system is now coursing through coaxial cables and beaming up and back from orbiting satellites.

The recent acceleration of speed to which our brains must adapt is staggering to contemplate. As I write this in 2023, the smart phone in our pocket has one million times more memory than the computer aboard Apollo 11 which landed the first human on the moon. And our smart phone has 100,000 times the processing speed.

Should we expect this new reality to be smooth, gentle, and easily adaptable?

One of the solutions for dealing with this new sped-up, fast-changing, digitalized age comes from educator Melissa Gouty, who addresses the paradox at the core of the digital screen in her article "Why You Need to Develop a Biliterate Brain—And How to Do It:":

> We live in a modern world where we daily navigate the vast network of digital sources. But we also have to be able to feel, understand and enjoy the depth of literature ... We need biliterate brains capable of both FAST and SLOW styles.
>
> Digitally-developed brains are capable of flipping through visuals, assimilating ideas, jumping from thought to thought, and finding information FAST. Print-oriented brains allow us to SLOW DOWN, savor the language, and consider the implications of the words on our lives.[75]

Note that the proposed "biliterate brain" connects closely to the previously seen distinction of Kathrine Hayles between "hyper attention" and "deep attention."

There are also practical reasons associated with the shift from printed page to digital screen: Melissa Gouty, a career educator, who clearly loves books and literature, nonetheless understands the economic and environmental advantages of the digital screen, noting that digital books cost less than printed books, so schools and libraries can purchase new inventory at a much lower cost. Readers can have access to materials through free online sites, which would not have been available to them if they had to pay for print books. She goes on to say that, since digital resources don't take up space, they don't necessitate building additional wings, constructing shelves, or figuring out where to house collections. Also, they are not subject to mold or mildew. Not only that, but one can access online materials at any time of day or night, and they are delivered within seconds. They are also portable and weightless.[76]

Given the ominous consequences of fossil fuel-generated technologies eroding the very ecosystem we need to survive on this planet, the environmental advantages of the digital screen over massive printing are clear. While acknowledging that there are environmental issues caused by the building and discarding of computers and computer networks, they pale in comparison to the horrific consequences of deforestation and fossil fuel toxicity created by the massive use of printed paper.

Consider this: "Over 15.2 billion pounds of newspapers and 2.5 million pounds of magazines were generated in the United States in 2014. Newspapers and magazines have a limited shelf life so switching to digital versions is a green thing to do."[77]

As much as I miss the tactile enjoyment of leisurely turning the pages of the voluminous Sunday *New York Times* (I now subscribe to the digital service), the environmental costs of just the Sunday newspapers are staggering: to produce all Sunday editions of newspapers requires cutting down 500,000 trees each week. That's 26 million trees, 126 billion gallons of wastewater and 73 billion pounds of greenhouse gases every week.[78]

9

NOW APPEARING *on* OUR DIGITAL SCREEN: THE COLLECTIVE UNCONSCIOUS

As Jung observed, "archetypes"— those images percolating up from the collective unconscious to appear in enduring myths, fairy tales, movies, TV, and quite extensively, in advertising campaigns, connect each of us to the history of thoughts, perceptions, beliefs, intuitions, and actions of all humankind past and present.

These archetypes, as well as all aspects of both our personal and collective unconscious, are responsible for almost every thought, feeling, belief, and action we experience. They can be seen as correlating to the World Wide Web, where we instantaneously encounter the self-organizing feedback system of thoughts, feelings, beliefs, and actions from the vast reservoir of humanity's knowledge and wisdom, as well as encountering our fears, anxieties, and deep concerns.

Keeping in mind the statue of Shiva, archetypal symbol of both the creative and destructive forces of the Universe, standing above the underground laboratory where the protocols creating the Web were first conceived, and where scientists continue to expand upon Einstein's dream of sledding near the speed of light, colliding particles at fantastic speeds in order to peer more deeply into the darkness of the quantum realm, here is a summation and some reflections:

- Quantum Physics, through wave-particle duality, the Uncertainty Principle, and Entanglement, revealed the world cannot be fully understood as separate objects but that, in the words of physicist Fritjof Capra, "Ultimately—as quantum physics showed so dramatically—there are no parts at all. What we call a part is merely a pattern in an inseparable web of relationships."

- Marshall McLuhan, in the 1960s and 1970s noted that all significant technologies are extensions of the human body: the wheel an extension of the foot, clothing an extension of the skin, the telescope and microscope extensions of the eye, electronic media an extension of the human nervous system. And while he died in 1980, just as the first personal computers were starting to appear in the mass market, he anticipated their becoming the extension of our human brain, leading to Nobel-Prize winning Stephen Hawking's statement, "We are all now connected by the Internet, like neurons in a giant brain."

- The emergence of the World Wide Web in the last decade leading up to the New Millennium and expanding exponentially as the twenty-first century unfolds, is not a coincidence. It is a reflection of **the Quantum & the Dream** synchronicity in the first year of the Modern Age, 1900, which can now be considered the first clarion call to humanity described by historian Thomas McFarlane: "In the 20th century the modern materialistic world view began to unravel in the face of scientific and psychological developments. It led a number of thinkers to consider that the human psyche may be more involved, in some mysterious way, with the observed properties of matter."

This clarion call is perfectly summarized by neuroscientist Shaikat Hossain, who writes:

"Some have learned how to adapt to the 'always on' mindset that the Internet imbues, whereas others have become distanced and long for the intimacy of offline interaction. It is within this burgeoning cloud of techno-social growth that the unique opportunity to study the human mind at large has presented itself."[79]

Referring to the World Wide Web, journalist/artist Ephrat Livni points out that

".... whether we judge it to be positive or negative, [it] is simply a manifestation of ourselves. It's the human psyche made manifest, with the same psychological qualities we ourselves possess."[80]

10

BUT WAIT—WHAT ABOUT ALL
that DAMN NOISE?!?!

An Estimated 330 Million People Will
Potentially Suffer from Internet Addiction
in 2022.
—INFLUENCER MARKETING HUB

Since Shiva, the archetype of the collective unconscious, reflects both the creative and destructive aspects of Nature, let's look more closely at the potentially destructive consequences of the World Wide Web.

Every day we hear loud, legitimate complaints of how frustrating and disappointing the World Wide Web is, saturated with X, formerly known as Twitter, feuds, fake news, corporate surveillance, cyber-attacks, identity theft, bullying, scams, and incessant pop up and banner ads clawing for our attention at speeds never before experienced by the human brain.

And what about the growing problem of people getting addicted to the digital screen, particularly many younger users? According to the Influencer Marketing Hub,

> Social media platforms such as Facebook, Snapchat, and Instagram produce the same neural circuitry caused by gambling and recreational drugs. A constant stream of retweets, likes, and shares affects the brain's reward area and triggers the same kind of chemical reaction as other drugs, such as cocaine.

> Some neuroscientists have compared social media interaction to a syringe of dopamine being injected straight into the system . . . an estimated 330 million people will potentially suffer from Internet Addiction in 2022.[81]

Tim Berners-Lee's "dream" of a World Wide Web has within it the nightmare of billions of human brains spewing out the worst of human fear, and greed as giant corporations continue a feeding frenzy for controlling the flow of information.

Before countering the torrid stream of criticism and warnings of impending doom being heaped on the World Wide Web, here is some more evidence of the destructive nature of the Web from critics.

In October of 2021, Frances Haugen, a former Facebook employee, said in a TV interview that Facebook realized that if they change the algorithm to be safer, people will spend less time on the site and click on fewer ads, so they'll make less money.[82]

She revealed how Facebook was being used in some countries to increase human trafficking, expand drug cartels, and promote violence while the executives at Facebook consider this "simply the cost of doing business."

As will be shown in the next chapter, the same addictive pitfalls have been true for all new media technologies throughout human history. It's also true that all emerging new media have expanded the flow of information and the ability of humans to socialize, a key factor in the rise of homo sapiens as the most powerful species on the planet.

We are reminded by Carl Jung, "The psyche is far from being a homogenous unit – on the contrary, it is a boiling cauldron of contradictory impulses, inhibitions, and affects."[83]

In her opinion column for the *New York Times*, "Why People Are So Awful Online," Roxanne Gay captures the frustration aimed at the World Wide Web:

It is infuriating. It is also entirely understandable. Some days, as I am reading the news, I feel as if I am drowning. I think most of us do . . .

After a while, the lines blur, and it's not at all clear what friend or foe look like, or how we as humans should interact in this place. After being on the receiving end of enough aggression, everything starts to feel like an attack. Your skin thins until you have no defenses left. It becomes harder and harder to distinguish good-faith criticism from pettiness or cruelty. It becomes harder to disinvest from pointless arguments that have nothing at all to do with you. An experience that was once charming and fun becomes stressful and largely unpleasant. I don't think I'm alone in feeling this way. We have all become hammers in search of nails.[84]

QUESTION: IS THIS AN INHERENT FLAW IN THE WORLD WIDE WEB OR IN US, THE USERS?

Internet technology with the communications protocols of the Web has become an extension of human consciousness (and even more so, the unconscious). It gives access to billions around the world both the historic record of knowledge and wisdom AND a two-way, participatory medium in which we all necessarily participate in the consequences of our every action. Adjustments, as with any new major media technology, take time.

Every new significant technology, particularly new media, has brought with it both the ability to expand perception and knowledge AND to increase confusion, disequilibrium, addiction, and anxiety. As Harvard Law Professor Noah Feldman points out, it's cheaper to produce lies rather than truth, so a lot of false information proliferates in the marketplace. We have a tendency to believe what we hear even when we shouldn't, especially when those things seem to confirm our prior beliefs.[85]

The negative effects of the Web are more pronounced than any media preceding it because of its globally connected power. But were we really better off some seventy years ago when the news and its context were controlled by a handful of corporations?

THOUGHT EXPERIMENT:

Imagine we could travel back one hundred years to the annual Solvay Conference, where Albert Einstein, Niels Bohr, Werner Heisenberg, Wolfgang Pauli, and the leading explorers of quantum theory met to converse about not only the significant scientific results but the philosophical implications of the fascinating, novel, unpredictable revelations of Nature being revealed. Imagine telling them that one hundred years in the future there would be a portable screen through which they could talk, correspond, and share insights without leaving their homes. And that, with the ease of a few keystrokes, they could access virtually the sum total of scientific, philosophical, and spiritual insights handed down through the ages. Would they not be convinced that this would lead to a startlingly expansive New Renaissance, despite the negative consequences of corporate manipulation, disorienting floods of information, and bad actors clawing for attention?

To gain a fuller understanding of what's taking place in the age of the World Wide Web, next on our journey through **the Quantum & the Dream** we'll explore more deeply the underlying, unconscious process of change inherent in the appearance of all new media throughout human history.

As we descend further into the underlying patterns of **the**

Quantum & the Dream, here is a useful insight from popular historian/speaker Yuval Noah Harari:

> "For thousands of years, we have gained the power to control the world outside us but not to control the world inside."[86]

BLASTS *from the* PAST:
GET RID *of* TV, BOOKS, *and the* ALPHABET

Throughout history, people have always worried
about new technologies. . . . The human brain
is always dealing with a constant stream of rich
information - that's what the real world is.
—DEAN BURNETT

When the television screen became standard in virtually every US home, there were screams about the scourge of addiction.

As Elizabeth Hartney wrote on the website *VeryWellMind,* "Although early research into TV addiction was limited, the concept of TV addiction was relatively well accepted by parents, educators, and journalists, as television watching became more common, particularly among children."[87]

A study published by the American Psychological Association concluded that early childhood exposure to TV violence predicted aggressive behavior for both males and females in adulthood.[88]

The same heightened concerns as those about the growing influence of television were present in the fifteenth century, when the printing press emerged during the Northern European Renaissance. Carl Gessner, a respected fifteenth-century Swiss scientist, may have been the first to sound an alarm about the effects of information overload. In a landmark book, Gessner described how the "modern world" overwhelmed people with an overabundance of data that was both "confusing and harmful" to the mind, a warning referring to the "seemingly unmanageable flood of information" unleashed by the printing press.

Neuropsychologist Vaughan Bell writes, "Worries about information overload are as old as information itself, with each generation reimagining the dangerous impacts of technology on mind and brain. From a historical perspective, what strikes home is not the evolution of these social concerns, but their similarity from one century to the next, to the point where they arrive anew with little having changed except the label."[89]

Now, here in the twenty-first century with the power of computer intelligence expanding the digitally-connected environment at speeds never before experienced by the human brain, the anxiety levels inherent with the arrival of any new media throughout human history are ramping up at astonishing levels.

The arrival of the personal computer as a major communications tool—and altering the sense ratios of the human brain—was only about forty years ago, yet now over five billion of us are connected through computers into a totally new digital, globalized world, causing explosive rupturing of traditional forms of education, economic jobs, and other cultural forms.

In 2021, Farhad Manjoo wrote in the *New York Times*: "We — the news media in particular and society generally — may be tripping into a trap that has gotten us again and again: A moral panic in which we draw broad, alarming conclusions about the hidden dangers of novel forms of media, new technologies or new ideas spreading among the youth."[90] As previously noted, Plato, one of the most revered philosophers to this day, had great concerns about the creation of the alphabet—it would erode the oral tradition and shift attention from Nature's voice to artificially constructed symbols.

We can agree that in the age of the digital screen, the "constant stream of rich information" becomes a raging, debilitating FLOOD if we don't adapt to it, including the need to take ample time away from the digital screen, re-orienting ourselves in nature, spending more quiet time in personal contemplation and inner reflection. Our right hemisphere is the gift from evolution for both understanding the unpredictable, potentially enlightened messages from our unconscious and for opening up to the boundary-expanding, uncertain, potentially enlightened new connections available through the globally hyper-linked Web.

Depth psychologist Thomas Moore, in his *Care of the Soul*: A *Guide for Cultivating Depth and Sacredness in Everyday Life*, cites the crucial need to take a more right-hemisphere, "big picture" view of where we are culturally and psychologically in the new, explosive, digitally-connected world:

> One day I would like to make up my own DSM (Diagnostic and Statistical Manual for Mental Disorders) with a list of "disorders" I have seen in my practice. For example, I would want to include the diagnosis "psychological modernism," an uncritical acceptance of the values of the modern world. It includes blind faith in technology, inordinate attachment to material gadgets and conveniences, uncritical acceptance of the march of scientific progress, devotion to the electronic media, and a life-style dictated by advertising.[91]

From the "big picture" view of **the Quantum & the Dream**, the immense advantages of the digital screen—and, at the same time, its addictive power to overwhelm the senses and blind us to the depths of our being—are reflected in the mythic, symbolic nature of the statue of Shiva, Creator and Destroyer of the Universe, standing on the grounds of CERN, the world's largest particle accelerator, where Tim Berners-Lee created the protocols leading to the World

Wide Web and where many of the world's leading scientists continue to probe the inner workings of the quantum realm.

Shiva reflects the deep understanding that creation and destruction are not separate, clearly defined actions but are, like the principle of yin-yang, two opposite poles of a unified energy source reflecting the subatomic dance of constant flux and creative transformation into new forms. "Renaissance" literally means rebirth, and Shiva reminds us that for a renaissance to emerge, older forms of perception and understanding have to die.

12

A WORD *from the* WEB CREATOR
& *the* MEDIA PROPHET

As we crossed the threshold into the New Millennium, Tim Berners-Lee, who single-handedly developed the protocols which created the World Wide Web, voiced disappointment that his dream of the WEB has been tarnished by misuse:

> While the web has created opportunity, given marginalized groups a voice, and made our daily lives easier, it has also created opportunity for scammers, given a voice to those who spread hatred, and made all kinds of crime easier to commit.
>
> Against the backdrop of news stories about how the web is misused, it's understandable that many people feel afraid and unsure if the web is really a force for good. But given how much the web has changed in the past 30 years, it would be defeatist and unimaginative to assume that the web as we know it can't be changed for the better in the next 30. If we give up on building a better web now, then the web will not have failed us. We will have failed the web.[92]

Marshall McLuhan, in his influential 1964 book *Understanding Media: The Extensions of Man,* offers a "big picture" view of where we were headed psychologically, writing that, after three thousand years of fragmentary and mechanical technologies—an explosion—

now the Western world is imploding. Our central nervous system has extended itself in a global embrace, effectively "abolishing both space and time as far as our planet is concerned."[93]

He went on to predict, "Rapidly, we approach the final phase of the extension of man— the technological simulation of consciousness, when the creative process of knowing will be collectively and corporately extended to the whole of human society, much as we have already extended our senses and nerves by the various media."[94]

Could this really be what's going on under the surface of the twenty-four-hour news cycle?

13

WELCOME *to the* THEATER *of* SIGNAL-*to*-NOISE RATIO: BEWARE *the* "URGENT BLATHER"

We have clear phone lines and Internet connections because scientists, most notably Claude Shannon, the "father of information theory," figured out how to reduce noise in transmission lines and increase clear signals. Shannon's formula, a mathematical theory of communications, is called "signal-to-noise ratio."

With the astounding achievement of linking billions of human brains together into a globally-connected WEB transmitted through the digital screens of our laptops, smart phones, and desktop computers, modern communications signals are technologically clear. But now it's our challenge to navigate through the flood of information noise to tune into the signals of knowledge, insight, wisdom, and connectedness now available.

There's an analogy between the signal-to-noise ratio from the physical/mathematical science of providing clear signals to the psychological/spiritual lens of **the Quantum & the Dream**: whenever any of us shifts to the right brain hemisphere to contemplate, meditate, create a thought experiment, or take off on a flight of imagination, don't we have to cut through the incessant noise of our chattering mind to get to the more enlightened, revelatory signals below?

This is why contemplatives and creators of thought experiments for thousands of years, before and during the arrival of sophisticated, ever-expanding new media technologies, have been developing inner mind techniques such as focusing on the breath, sound mantras, and visualization guides: to slow down the brain waves and reduce the noise of the stressed, anxious voices we inevitably contact when turning the camera of consciousness inward.

Novelist Tim Parks, writing in *The New York Review of Books*, notes that, "By far the main protagonist of twentieth century literature must be the chattering mind, which usually means the mind that can't make up its mind, the mind postponing action in indecision and, if we're lucky, poetry."[95]

He points to Hamlet's suffering from the dissonant voice inside his head, inducing a paralyzing indecision, and to the observation of the great writer Virginia Woolf on the experience of looking inwards to the deeper layers of unconscious wisdom, where the brain "sounds darker notes, warning us that the mind risks being submerged by the urgent blather of modern life."

The modern industry that researches, understands, and takes advantage of the human brain's constant, anxious "chatter," delivering finely-tuned messages to offer the anxious mind comfort and solutions, is clearly advertising and marketing.

Every major new communications technology becomes a bigger platform for both progress and for greed and deception. One of my favorite examples is the first stock exchange market which opened in the Dutch city of Amsterdam in 1602.

At the time this first stock market opened, it was rightly seen as a novel, progressive medium through which, for the first time, citizens would have the opportunity to invest in large companies.

Within days of its opening, a founder of the exchange, realizing the potential profit of circulating "fake" news, made it public he was selling a big part of his portfolio knowing it would cause the price of those stocks to drop. He then bought them back at the lower prices. Soon a number of fellow brokers, seeing the success of the scheme, joined in the stock manipulation game just days after the initial opening!

Fast forward to the year 1930 when newspapers were the main

advertising medium, along with the new medium of radio. An event called "Torches of Freedom" was the most anticipated highlight of that year's Easter Parade down New York City's Fifth Avenue.

"Torches of Freedom" sounds like a campaign with a high-minded ideal.

In fact, it was a public relations stunt to get women to smoke more cigarettes.

The man behind the event, Edward Bernays, became known as "The Father of Public Relations." He has had as much influence on American culture as just about any individual, given how his strategies, based on a keen understanding of how to manipulate the human unconscious, have grown into the modern advertising industry.

Bernays writes that his "Torches of Freedom" campaign was financed by the American Tobacco Company: "Hill [the President of the American Tobacco Company] called me in. 'How can we get women to smoke on the street? They're smoking indoors. But, damn it, if they spend half the time outdoors and if we can get 'em to smoke outdoors, we'll damn near double our female market. Do something. Act!'"[96]

The marketing ploy was hugely successful, significantly boosting cigarette sales to women.

One popular advertising poster showed a beautiful woman gazing beatifically up at a message in large letters reading, "Believe in Yourself." Economically brilliant, ethically bankrupt, Bernays' strategy was to link smoking outdoors to women's rights to freedom of expression.

And this successful campaign occurred before the existence of TV and the World Wide Web.

In the mid-1950s, like millions of Baby Boomers, I spent hundreds of hours as a child hypnotized by the relatively new invention of television, eyes glued to the screen as advertisers successfully launched a constant stream of cleverly produced, manipulative ads for sugar-laden, chemically processed cereals, cakes, and candies along with a constant parade of enticing toys which immediately mesmerized my young brain, creating the desire for material items I just had to have.

While the World Wide Web greatly amplifies these issues, it also offers a crucial advantage—none of us could directly talk back at those ads projecting through the TV screen and into the addictive parts of our brain. With the World Wide Web we now have a two-way medium through which we can talk back.

Now if someone wants to get good critical insight into the worth of something being sold, ample consumer opinion is just keystrokes away through the digital screen.

Journalist Roxanne Gay quoted above referring to the "drowning" sensation so many feel from the "infuriating" over-commercialized noise of the WEB makes the point, "At least online, we can use our voices and know they can be heard by someone."

The calling of **the Quantum & the Dream** synchronicity for the current zeitgeist is that the heightened anxiety we feel now is as much from the "blather" inside our own heads as it is from the incessant echo chamber of Facebook posts, Instagram feeds, TikTok videos, and X (formerly known as Twitter) feuds, narcissistic polemicists, and money-driven corporate interests.

Here it is useful to reflect again on the basic differences between the two hemispheres in the frontal cortex of our brains, since evolution has provided us with different inherent proclivities:

The left hemisphere, wired for sharp focus on details and the ability to "grasp" for objects, was essential for our hunter-gatherer ancestors: They had only primitive weapons to enable survival in an environment of saber-toothed tigers and other predators which were much faster and stronger than humans.

But as Homo sapiens gained technological mastery of tools and weapons, this same left hemisphere, wired "to grasp" and control objects, but bereft of the "big picture" prescience of the right hemisphere, has led to our current crises. We've allowed ourselves to be programmed to "consume," to "grasp" for material objects as palliatives to ease the pressure of discovering the depth of our being (the subject of the next section of our trip).

As we speed ever faster through the current zeitgeist, we need a shift away from the "grasping" instincts of a left-hemisphere perception, which sees the world as separate, material objects, and which embodies the insatiable need for more material resources. In order to take advantage of the incredible benefits of the globally-hyperlinked Web on both the personal and collective/societal levels, now, more than ever before in human evolution, we require a shift to the more emotionally empathic, imaginative, "big picture" potential inherent within the right hemisphere of our brains.

The erosion of mortar-and-brick factory style/assembly line schools resulting from the explosive influence of the digital Web can be traced back to the discovery of the quantum and the cultural interest in the unconscious.

As noted earlier, philosopher and scientist Thomas McFarlane observed that the modern materialistic worldview began to unravel in the face of psychological and scientific developments, and led many thinkers to consider that the human psyche may be mysteriously involved with the observed properties of matter.[97]

If you have any doubts about the unraveling of the materialist paradigm at the core of **the Quantum & the Dream** synchronicity, fasten your seat belts for what comes next.

As we descend deeper into the underlying patterns of evolutionary shift taking place—as the debilitating noise of the 24/7 news cycle gets even louder, as the exponential advances of computer intelligence permeate more and more of our daily life, as anxiety levels continue to accelerate—can we keep track of the deeper signals of transformation spiraling up from the unconscious depths for both our personal and collective benefit? Remember Shiva continuing its mythic dance of creation and destruction above the world's highest energy particle accelerator probing the subatomic quantum realm.

As we prepare for the final leg of our journey through **the**

Quantum & the Dream, another right-hemisphere, "big picture" insight from Marshall McLuhan, "The Oracle of the Electronic Age," is particularly relevant: "There is no such thing as inevitability as long as we are willing to contemplate what is happening."

ORIENTATION *for the* NEXT LEG *of the* JOURNEY: APOCALYPSE NOW?

Turning and turning in the widening gyre
The falcon cannot hear the falconer;
Things fall apart; the centre cannot hold . . .
—W. B. YEATS

A Zen master enters the studio of my radio talk show.

John Daido Loori is a world-renowned Zen teacher, naturalist, author of over twenty books, and founder of the Zen Mountain Monastery near Woodstock, NY.

I ask Daido: "Can you give us a description of a personal enlightenment experience?"

He tells of arranging a ride in an official National Oceanic and Atmospheric Administration Hurricane Hunter, a specially built plane for flying through hurricane force winds into the eye of the storm, gathering important data which can save lives.

Daido talks about how the trained Hurricane Hunter pilot, like a Zen practitioner, can only learn how to maneuver in such violent winds by actually flying through them—no amount of ground training can simulate the split-second, visceral instincts needed to fly a plane in such unpredictable, intimidating conditions.

Daido recounts the anxiety he felt as the plane was being bounced around by one-hundred-plus mile-per-hour winds, the sky filling with sudden bursts of lightning ... then finally breaking through the darkness of the eyewall of the storm into the bright center of the eye, a wide, cylindrical tube which is calm and brightly illuminated, offering a wide view in all directions.

Fast forward just over a year. My in-studio radio guest is renowned dream teacher Dr. Jeremy Taylor, a cofounder of The International Association for the Study of Dreams.

The topic of discussion is **nightmares**. Jeremy explains how, after decades teaching group dream work across the US and in a number of foreign countries, he saw ample evidence that there's no such thing as a bad dream.

In fact, often the most overwhelming nightmares are a calling from our inner psyche, saying that it's time to pay attention to something really important! Even the most gut-wrenching, heart-palpitating nightmare is a call from the unconscious that a big issue needs to be resolved. If we are willing to replay the dream, imagining and taking note of the most striking images, there's a good chance the deeper message of the nightmare will emerge, sometimes through a synchronistic experience.

Jeremy's favorite historic example of a terrifying nightmare bringing an enlightened message is that of Elias Howe.

In 1846, as the Industrial Revolution was just gaining momentum, there was a huge call for a more efficient sewing machine. Despite months of working long, tedious hours, Howe was no closer to a solution, and getting enormously frustrated. Falling asleep on his workbench one night he experienced this nightmare:

> In his dream he was fleeing from cannibals through the African jungles. Despite his frantic efforts, he was unable to escape them. They captured him, bound him, and took him back to their village where they tossed him into a huge pot of water to cook alive. As the water began to bubble up around him, his bonds loosened and he was able to free his hands. He

attempted to climb out of the pot, but each time he clambered up over the edge, the natives would reach over the flames and poke him back into the cauldron with their sharp spears.[98]

He wakes up from the nightmare sweating and in emotional turmoil. Yet an intuitive part of his brain shifts his awareness to something odd about those sharp spears poking him back into the boiling water—there were holes near the point of every spear.

"Holes in the points." The image brought forth an "AHA!" of recognition. As with others seeking to design a more efficient sewing machine, Howe had assumed that the hole in the machine's needle should be in the same place as a handheld needle—at the base. But Howe realized the nightmare came to show him that the point of the needle for the machine needed to be moved from base towards the pointed end.

He quickly redesigned the gear system with the reconfigured needles, which allowed the needle to push thread through layers of cloth and pull up again with great speed and efficiency. Howe's invention is credited by historians as a key generator of the Industrial Revolution. And unless you're now reading this either stark naked or in clothing completely hand-sewn, the clothes you're wearing are a direct result of Howe's terrifying nightmare.

Jeremy loved to emphasize the key to Howe's discovery: Howe wasn't a dream expert, but he was able to move through the emotional turmoil of the terrifying nightmare and note how his intuition sought out the key image of the dream. Howe's nightmare also reveals how our unconscious loves to communicate with puns and a sense of play as the "point" of the sharp cannibal spears "pointed" the dreamer to the dream's most important "point."

We can see the connection between Zen Master Daido's experience flying through the shattering turbulence of a violent hurricane into the calm, illuminated eye of the storm and the experience of Elias Howe seeing through the emotional turmoil of a terrifying nightmare to discover the enlightened message contained within.

This connection serves as a useful guide to the next part of our trip through **the Quantum & the Dream.** For we are all coming face-to-face with an apocalyptic, collective nightmare: the potential collapse of the very biosphere responsible for advanced life on this planet.

In *Dream Work: Techniques for Discovering the Creative Power in Dreams,* Jeremy Taylor refers to the important insight of psychiatrist Sandor Ferenczi: "The dream state is the workshop of evolution."

ARE WE HEADED TOWARDS MASS ANXIETY OR THE NEW RENAISSANCE? PROBABLY BOTH.

Here in the third decade of the twenty-first century, mass anxiety is already well under way. While there is no certainty that we will bring the New Renaissance to fruition, there are similar patterns emerging which echo those that helped to induce the last great renaissance during the fourteenth to sixteenth centuries.

The causes involved in the creation of any renaissance period are complex, but they are always accompanied by a seismic, systematic breakdown. In the case of the Italian Renaissance, the new, enlightened "humanist" philosophy was in large part a reaction to the devastation of the Black Plague, which ended up wiping out about one third of the entire population of Europe. The plague broke down societies and at the same time inspired the greatest flourishing of philosophy, art, and science in human history.[99]

Between 2019 and 2023, COVID-19 and its variants, which caused the unprecedented shutdown of the global economy, killed close to seven million people worldwide.

Nonetheless, this virus pales in comparison to the devastation of the Bubonic Plague. It is the growing, catastrophic effects of what we politely refer to as climate change (a more accurate description being ecological collapse), that may be recapitulating the dynamic of societal collapse from Black Plague, a key pressure leading to the emergence of the paradigm-shifting Italian Renaissance and its spread northward into western Europe.

As I write this in 2024, here is a brief sketch of the violent storm of environmental collapse already occurring, as well as evidence of mass anxiety permeating the planet. Contemplating the impeding storm, it's useful to remember Zen Master Daido's harrowing trip into the calm, illuminated eye of the hurricane and Elias Howe's enlightened discovery by contemplating an image from his horrific nightmare:

- On September 18, 2021, Secretary General of the United Nations, Antonio Guterres, pointing to a lack of effort by leading countries to deal with "deadly" heat waves and rising sea levels said, "We're digging our own graves by burning, drilling, mining deeper. . . . We are heading for climate disaster. Young people know it. Every country sees it."[100]

- Carbon dioxide from human activity is increasing more than 250 times faster than it did from natural sources after the last Ice Age.[101]

- Air pollution destroys billions of years of life expectancy: millions of people die each year from air pollution levels exceeding international guidelines.[102]

  An important perspective often lost: While COVID-19 and variants had the devastating effect of killing some seven million people over three years—in comparison, air pollution is killing over 6.6 million people EVERY YEAR![103]

- The extinction of plants and animals is accelerating one thousand times faster than before humans showed up on the planet. Nearly three billion birds have been lost in North America since 1970, causing a huge drop in pollination rates which means far fewer new natural food resources.[104]

- Bees are crucial contributors to the pollinating plants and the growth of crops. Yet, researchers have recently determined that the American bumblebee population is at near extinction levels, having declined by 89 percent over the last twenty years. This could greatly impact crop production and impact whole ecosystems.[105]

- The consequences of overpopulation cannot be overstated. In 1960 there were around three billion people on Earth. In November 2022, the number hit eight billion! Every day that passes, the numbers grow and the collective human impact on the environment intensifies. We are literally causing a loss of breathable space. It's simple: the more humans living on the planet, the less room for any other life. The major driving force behind the loss of ecosystems—rainforests, coral reefs, wetlands, and Artic ice—is human overpopulation. Rainforests once covered 14 percent of the planet's land surface. Now they cover barely 6 percent, and experts estimate that the remaining rainforests could be consumed in less than forty years.[106]

- Rainforests are crucial for absorbing carbon dioxide and releasing oxygen, a process on which our survival depends—yet *half of the world's rainforest has been destroyed in just one century.*[107]

- A study by the UNEP Global Environmental Outlook, involving 1,400 scientists and which took five years to prepare, found that human consumption has far outstripped available resources. Finite resources, like fossil fuels, fresh water, arable land, coral reefs, and frontier forests, are plummeting, placing competitive stress on the resources needed to sustain life. Each person on the planet now requires a third more land to supply their needs that the earth can supply.[108]

- Journalist Graham Peoples observed in 2021, "Excessive consumerism is the wind driving greenhouse gases into the atmosphere, and the poor consume little. It's the rich that greedily devour (multiple houses, boats, cars, private jets/ air travel, loads of stuff) followed by the consuming masses in developed countries. Conspicuous consumerism is a man-made disease; an artificial ideologically induced virus enflamed by desire."[109]

- Over-consumption and overpopulation are depleting the earth of fresh water, one of the most essential resources for life. "70 percent of freshwater is icecaps, and the remaining 30 percent makes up land surface water such as rivers and lakes. Most of the freshwater resources are either unreachable or too polluted, leaving less than 1 percent of the world's freshwater, or about 0.003 percent of all water on Earth, readily accessible for direct human use."[110]

- In the US, the wealthiest country on the planet, one of its primary sources of water, the Colorado River, is disappearing. According to a *NY Times* headline, in August 2021, "40 Million People Rely on the Colorado River. It's Drying Up Fast."[111]

- And an irony too dramatic to ignore, as freshwater supplies dangerously shrink, the opposite—increased flooding—

wreaks havoc. While it's true that sea levels rose steadily over the last century, the rate in the last two decades is nearly double that of the last century and accelerating every year.

Nearly 40 percent of the United States population lives in coastal areas, meaning tens of millions of people will likely suffer from the rising sea levels.

In the summer of 2021 Europe suffered massive, deadly floods from intense rainfall, destroying billions of dollars of property and taking hundreds of lives. As reported by CNN: "deadly floods in Germany, Belgium and the Netherlands were up to 9 times more likely because of climate change."[112]

Floods destroy food sources and transportation routes. Also, any country that regularly experiences flooding expends money on recovery instead of growth and development. Livelihoods are disrupted and businesses are understandably unwilling to invest in disaster-prone areas, so development of these areas will be at a standstill.[113]

- Increasing numbers of wildfires, intensified by drought and climate change, are destroying millions of acres of land and polluting millions more. Wildfires are the largest source of deadly air pollution in California. In 2021, California fires and high ozone levels turned the air in Denver and Salt Lake City into some of the dirtiest in the world, worse than New Delhi's or Beijing's.[114]

- As reported in January, 2022 due to Global Warming, "Victoria, New South Wales, Queensland, and Western Australia have been ravaged by bushfires that have so far taken the lives of an estimated billion animals. The blazes now encompass an area equivalent to the size of Denmark and Belgium combined, leading to massive evacuations of residents and tourists."[115]

- According to a report by the Intergovernmental Panel on Climate Change published by the United Nations, a staggering 143 million people will likely be uprooted over the next thirty years by rising seas, drought, searing temperatures, and other climate catastrophes.[116]

- No country is being spared. The ten hottest years worldwide on record have occurred since 2005.

As we move through the turbulent headwinds of climate collapse, the effects on the human psyche, as evidenced by humanity's nervous system on high alert, is growing. And it started well before the devastating effects of the COVID-19 pandemic.

- In 2017, before the COVID-19 global shutdown, The World Health Organization (WHO) found almost one out of every five persons worldwide suffered from medically confirmed anxiety disorders, the three most common being specific phobias, major depressive disorder, and social phobias.[117]

- It was also reported that anxiety disorders are the most common mental illness in the US, affecting forty million adults age eighteen and older.[118]

- According to a report published in the Journal of the American Medical Association, "Mental health-related visits to emergency rooms by children, teenagers and young adults soared from 2011 to 2020.... The sharpest increase was for suicide-related visits, which rose fivefold. The findings indicated an 'urgent' need for expanded crisis services, according to the team of researchers and physicians who published the report."[119]

- While developing countries in Africa, and poorer countries in Asia, Central America, and South America will feel the brunt of increasing ecological collapse, wealthy countries will suffer greatly as well from increasing economic stresses: According to 2016 data, nearly half of US households headed by someone fifty-five or older have no retirement savings. Many Americans over sixty-five face trying to get by on Social Security income alone, which provides an average retirement benefit of $18,516 a year.[120]

- Compare that figure with the cost of long-term health care. According to the 2020 Genworth Cost of Care Survey, the median cost of a semiprivate room in a nursing home is more than $93,000 a year, and the median cost of employing a home health aide full time is around $50,000 per year.[121]

So, within all this frightening physical, emotional, and spiritual turbulence, what might the calmer, illuminated eye of the storm look like? Can we get there?

In the next section, we will look at evidence, not immediately obvious in the din of the twenty-four-hour news cycle and growing anxiety around the globe, that there are in fact seeds already planted that could blossom into the New Renaissance.

THIRD SHIFT
FROM THE AGE OF INFORMATION
TO THE AGE OF RECOGNITION

The human eye sees the physical form, but the inner eye penetrates more profoundly, even to the universal pattern of which each person is an integral and individual part.

—SRI YUKTESHWAR GIRI

1

THE ULTIMATE GAME: EVOLUTIONARY LEAP

In the process of being broken open, worn down,
and reshaped, an uncommon tranquility can
follow. Our undoing is also our becoming.
—TERRY TEMPEST WILLIAMS

Here is a **Brief Kinaesthetic/Thought Experiment** that can illuminate the remainder of our journey:

Focus on the palms of both hands.

Feel how the five fingers on each hand extend outwards from the palm.

Slowly close each hand into a fist and feel your fingers pressing inwards, generating a concentrated pressure.

Now, if you have a pen, coffee cup or any small object nearby, pick it up and "grasp" it, paying attention to how the thumb's flexibility provides added support.

Put down whatever object you were grasping and slowly open your hands, palms facing you, letting your fingers extend outwards.

> Spend a few moments tuning into how this "opening up" of your hands changes your relationship to the rest of your body, and notice how you are looking outward into whatever is in front of you.

This simple experiment provides an understanding of what made our species the most powerful on the planet AND is now threatening not only our own existence, but that of the vast majority of life on Earth.

The grasping of the object you picked up connects back to the earlier discussion of how the left hemisphere of our brain is wired for "grasping," the right hemisphere wired more for "opening up" to a bigger picture of the world and how we fit into it.

The evolution of the grasping hand led to the development of sophisticated tools and advanced intelligence.

> The evolutionary ball started rolling, of course, when walking on two feet meant the hands were no longer needed for locomotion. They could then be used for a wide range of tasks: transporting food or offspring, scooping up water, gathering material to build a shelter or holding objects in one hand and manipulating them with the other to carry out specific tasks.[122]

As skillfully depicted in the movie *2001: A Space Odyssey*, the ability to grasp also produced weapons, such as the bone used by the ape-like leader of a tribe to bash the skull of another group's leader competing for space at a water hole.

In one of the most inventive and surprising jump-cuts in movie history, the ape-like animal, in a gesture of triumph over his kill, hurls the bone in the air, the camera follows the spiraling trajectory of the bone, which then morphs into a twenty-first century space shuttle carrying an American scientist. This scientist refuses to reveal information about a discovery on the moon to his scientific colleagues, connecting the grasping of a tool by the ape controlling the water hole to the grasping of information to control operations on the moon.

From our twenty-first century perspective, we recognize that, when restricted to the left-hemisphere's wiring for certainty and control, the same ability of our hands to anatomically grasp and develop the tools of advanced technology has led to the blindness of our over-heated, fossil-fueled desire for more and more material benefits, literally toxifying and destroying the biosphere responsible for advanced life on the planet.

2

WHAT'S *in a* NAME?

One of the most significant ways humans have tried to "organize" Nature, was to distinguish ourselves by naming every other species on the planet. We gave the name "Homo habilis," which means the tool maker, to our ape-like ancestors and then, when it came to naming ourselves, we chose "Homo sapiens," which means the "wise man." Not the "knowledgeable" man, but the "wise" man.

One of the key differences between "knowledge" and "wisdom" is context, the ability to create a bigger picture of how we best fit into the outside world. As seen on the journey through **the Quantum & the Dream** synchronicity, the "materialist paradigm" which has dominated human thought and action over the past five hundred years, may have brilliantly increased our knowledge, but clearly not our wisdom.

While it may appear rational to think we'll avoid the devastating potential of climate change by coming up with advanced technological tools to neutralize its effects, a reading of top experts in this field reveals that, while we would do well to focus on green energy alternatives to slow down the ongoing polluting of the planet, it's too late for any technological fix to reverse the process. Robin Kundis Craig, one of the leading environmental law scholars in the world, writes, "Geoengineering projects thus repeat the human hubris that has attended many much smaller-scale attempts to manipulate nature. Notably, however, this time the fate of the entire planet hangs intentionally in the balance."[123]

Thinking we can cure the ravages of global warming with better technology keeps us stuck in the left-hemisphere vision that we are separate from Nature and can manipulate it with our technology. This is confirmed by much research, such as this from the *Harvard Gazette*:

> Can technology save us from the worst effects of climate change? Probably not, reports a new study, 'Does Directed Innovation Mitigate Climate Damage? Evidence from US Agriculture,' published last month in the Quarterly Journal of Economics.
>
> The study showed that "we cannot simply innovate our way out of danger," said co-author Jacob Moscona, a Prize Fellow in Economics, History, and Politics at Harvard.[124]

What's needed, according to Professor Craig, is a new cultural story, that is, a new, more expansive vision of how to live within the seismic thrust of climate change:

> How we think about the natural world and our relationship to it matters. Moreover, these relationship stories are in fact a form of narrative—that is, a cultural story about how we exist with and within natural systems. . . . Narrative is a fundamental mode of human thought, and anthropologists have long studied creation stories, myths, folklore, and personal narratives for insights into how particular cultures construct and inform personal and cultural identity, give meaning to events.[125]

Journalist Adrienne LaFrance also advocates for a new cultural and philosophical vision, one which encourages us to turn our vision inward. In her article, "The Coming Humanist Renaissance" she writes, "In an age of anger, and snap reactions, and seemingly all-knowing AI, we should put more emphasis on contemplation as a way of being. We should embrace an unfinished state of thinking, the constant work of challenging our preconceived notions."[126]

LaFrance points to the paradoxical power of expanding AI to both advance humanity towards a renaissance and generate new levels of anxiety:

> Generative AI, just like search engines, telephones, and locomotives before it, will allow us to do things with levels of efficiency so profound, it will seem like magic. We may see whole categories of labor, and in some cases entire industries, wiped away with startling speed. The utopians among us will view this revolution as an opportunity to outsource busywork to machines for the higher purpose of human self-actualization. This new magic could indeed create more time to be spent on matters more deserving of our attention—deeper quests for knowledge, faster routes to scientific discovery, extra time for leisure and with loved ones. It may also lead to widespread unemployment and the loss of professional confidence as a more competent AI looks over our shoulder.[127]

Here in the twenty-first century, we find ourselves embroiled in an enigmatic grand game—a profound orchestration of evolutionary forces and temporal flux connecting us right back to that prehistoric moment when our ancestors, standing upright, with opposable thumbs, first grasped the conscious notion of tool/weapon.

This "closed-fist" awareness got us here. Now what?

More closed fist or open hand?

More information or more recognition?

The insight we previously looked at from philosopher/science historian Thomas MacFarlane, recognizes the deep cultural, philosophical, and psychological pattern we have been tracing: "In the

20th century the modern materialistic worldview began to unravel in the face of scientific and psychological developments. It led a number of thinkers to consider that the human psyche may be more involved, in some mysterious way, with the observed properties of matter."

We next travel back to the year 1900 to discover a surprising and empowering event which provides depth to **the Quantum & the Dream** synchronicity path we have been following.

3

A NEW HERO *for the* MODERN AGE

Our trip started out at the beginning of 1900 with Sigmund Freud's publication of *The Interpretation of Dreams,* and Max Planck's discovery of the quantum near the end of that year, setting in motion **the Quantum & the Dream** pattern, a descent into the illuminating power of the dream and into the mind-bending realm of the subatomic world.

Now we can add a third element to this synchronicity, emerging in that first year of the modern age, virtually midway between Freud and Planck's contribution. In May, 1900, a fictional character blew into cultural consciousness with the thrust of a gale force wind.

This character's storyline captured the public imagination to such an extent that the book sold out all ten thousand first print copies and went on to become one of the best-known and influential stories in American literature. In April, 2000, the Library of Congress called it "America's greatest and best-loved homegrown fairytale."

Directly connecting to Freud's book, this influential storyline is based on a dream, a dream which became even more embedded in the public psyche when, in 1939, it was made into one of the most popular movies of all time.

This popular book, appearing in 1900, midway between Freud's book on dreams and Planck's discovery of the quantum is *The Wonderful Wizard of Oz.* Its most unique and influential element is that, while it faithfully models what mythologist Joseph Campbell identifies as the "Hero's Journey," the main protagonist in this story is not the physically powerful, competitively-honed male figure at the center of virtually all previous heroic journeys. This hero, who risks bodily harm and death in the service of a soul-driven, inner transformation, is an adolescent girl: Dorothy Gale.

The powerful effect on the cultural psyche of a female-driven hero's journey cannot be overemphasized.

Author L. Frank Baum was surrounded by strong-willed, independently-minded women. His mother-in-law, Matilda Joslyn Gage, was an ardent supporter of women's suffrage and women's rights in general, coauthoring with Cady Stanton and Susan B. Anthony the three-volume *History of Woman's Suffrage*.

Baum's wife, Maude Gage Baum, was such a strong-willed, independent woman for her time that she became the central character of the successful 2019 novel *Finding Dorothy*, by Elizabeth Letts. Through her research for the book Letts commented, "I began to understand how Frank Baum wrote such remarkable female characters. A really new way of thinking about women's roles was percolating in the Baum household, and you can see reflections of it in Oz once you know to look for them."[128]

Letts was so impressed after first seeing *The Wizard of Oz* as a young girl that, for years after, she considered Dorothy to be her "invisible companion," a role model who later became the inspiration for her novel.

Dorothy's arrival was a game-changer. Her portrayal as a brave, independent, and resourceful protagonist offered a new archetype for modern female characters, inspiring a shift in the perception of women's capabilities as equal partners in the transformation of cultural values.

Dorothy's heroic trip down the Yellow Brick Road further connects to **the Quantum & the Dream** synchronicity, since Baum, like many of the key founders of quantum theory as well as Carl Jung, was deeply committed to the philosophical and spiritual insights of ancient Eastern mystics. Previous to publishing his *Oz* book, Baum wrote a series of editorials for a newspaper he edited, advocating the spiritual insights of Buddhism, Hinduism, the Tao, and other forms of spiritual wisdom. These taught that reality at its core—as with the unconscious mind and the inner matrix of quantum waves—is a non-material, non-linear, totally interconnected realm, one whose agent of change is not a ruler from above, but inherent drives from within.

The *Wizard of Oz*'s themes of self-identity, personal growth, and bravery in the face of great fear resonated powerfully with readers and viewers.

Dorothy's appearance in 1900, between Freud's publication and Planck's discovery of the quantum, adds even more strength to the call for expanding the limitations of the materialistic paradigm documented throughout our trip. As described by philosopher/physicist Fritjof Capra:

> The [materialist] paradigm that is now receding has dominated our culture for several hundred years...the view of the universe as a mechanical system . . . the view of the human body as a machine, the view of life as a competitive struggle...the belief in unlimited material progress . . . and—last, but not least—the belief that a society in which the female is everywhere subsumed under the male is one that is "natural." During recent decades all of these assumptions have been found severely limited and in need of radical revision.[129]

As a depth psychological journey, emphasizing the power of the dream as a tool for inner transformation, the *Oz* storyline and the immensely popular 1939 movie contained many of the powerful archetypes Carl Jung demonstrated as the core of the collective unconscious: the Warrior, the Good Mother, the Shadow, and the Trickster. As for the Warrior archetype, this time the hero protagonist, while demonstrating an indominable will to vanquish the enemy, also displays emotional vulnerability and the instinct that, through collaboration, one generates more strength than as a totally independent force. Note that all of the male characters in the story are physically and psychologically weak: Dorothy's Uncle Henry, incapable of standing up to the evil neighbor determined to take away Dorothy's beloved dog, the brain-challenged scarecrow, heart-missing tin man, cowardly lion, and charlatan wizard. Dorothy is their emotional, empathetic, integrative force as well as their valiant leader.

In addition to Dorothy, the most powerful figures in the story are the two female witches, one good, one evil, who create the tension for Dorothy's descent into the creative unconscious to find her true inner core.

To this day, Dorothy remains a valuable counter-balance to the cult of super-hero action figures. Her victory within the dream of Oz is more an inner, spiritually-oriented transformation than a rip-roaring action battle. Yet adolescent Dorothy clearly displays a strong, competent, and independent physical power.

The opening page of Baum's story describing the stark, arid Kansas farm that is Dorothy's home has an eerie connection to what's happening in many areas today due to the ravages of global warming:

> When Dorothy stood in the doorway and looked around, she could see nothing but the great gray prairie on every side. Not a tree nor a house broke the broad sweep of flat country that reached to the edge of the sky in all directions. The sun had baked the plowed land into a gray mass, with little cracks running through it. Even the grass was not green, for the sun had burned the tops of the long blades until they were the same gray color to be seen everywhere. Once the house had been painted, but the sun blistered the paint and the rains washed it away, and

now the house was as dull and gray as everything else.[130]

"Gray" is used as a descriptor four times in this description of the farmland, along with adjectives such as "baked," "cracks," "burned," "blistered," "dull."

The phrase "The sun had baked the plowed land into a gray mass," also anticipates the literary impact of John Steinbeck's Nobel prize-winning novel *The Grapes of Wrath*, whose opening is set in the American drought-generated dust bowl of the 1930s (yet another intriguing synchronicity: *The Grapes of Wrath* was published the same year as the movie release of *The Wizard of Oz*—1939).

Further connecting Steinbeck's dust-bowl novel to the "sunbaked land" of Dorothy's upbringing is the inherent strength of women adapting to turbulent change, as expressed in the following excerpt from *The Grapes of Wrath*:

> Women can change better'n a man," Ma said soothingly. "Woman got all her life in her arms. Man got it all in his head...Woman, it's all one flow, like a stream, little eddies, little waterfalls, but the river, it goes right on. Woman looks at it like that. We ain't gonna die out. People is goin' on-changin' a little, maybe, but goin' right on.[131]

The "sunbaked land with little cracks running through it" of Dorothy's upbringing echoes the ravages of global warming already wreaking havoc on farmland across the contemporary American prairie. And the tornado, a central force in both book and movie, is echoed in the 2021 CNN headline about destructive storms: "Climate change likely played a role in this weekend's deadly tornadoes."

THE ENIGMATIC SPIRAL OF DREAMS

Much has been written about the archetypal characters portrayed in both the original Oz book and the movie. Connecting the narrative to the "recognition" at the heart of this section of the journey through **the Quantum & the Dream**, it's interesting to look at the powerful symbol of Nature in the narrative.

Dorothy's initial call to adventure, her inner trip via dream, is generated by the violent winds of a tornado lifting her away into a colorful dreamscape. In fact, one of the key features of the 1939 movie was the switch from black and white scenes on the deprived farm to the glorious technicolor when she first sets foot in the dreamscape of Oz. In the realm of depth psychology and dream symbolism, a tornado often signifies the arrival of a profound life-changing transition, an indication that the dreamer's unconscious is signaling the potential for a new beginning in an important aspect of the dreamer's life.

While dreams of all types of storms have this potential meaning of renewal, the tornado is not only one of the most violent natural forces on this planet, it has the unique feature of containing Nature's (and, in fact, the entire Universe's) most influential shape: the *spiral*.

The spiral appears in entities as microscopic as the double helix of our DNA to the cosmic shape of galaxies. "The spiral is an ancient and mysterious symbol . . . the shape of a spiral is intrinsic to the makeup of the universe. . . . It's clear that spirals play a significant role in our lives, and ancient people knew this just as much as we do in modern times."[132]

In spiritual contexts, spirals often represent the continuous process of pushing outward while being part of a greater force.

Depth psychology, particularly the theories of Carl Jung, emphasizes the significance of the unconscious mind in shaping our thoughts, behaviors, and emotions. The spiral is an apt symbol for the journey into the depths of the unconscious, where the exploration and integration of repressed or forgotten aspects of the Self lead to greater self-understanding and psychological wholeness. Jung wrote: "We can hardly escape the feeling that the unconscious process moves spiral-wise round a center, gradually getting closer, while the characteristics of the center grow more and more distinct."[133]

The spiraling motion connects us directly to the evolutionary pattern of Nature: our being alive depends on our hearts pumping blood throughout our body in a spiral vortex. "For centuries, artists, biologists and mathematicians have been inspired by the recurring patterns of the plant world: the exquisite symmetry of flowers, the sweeping spirals of seeds, spines and leaves."[134]

Further connecting the twisting spiral Oz tornado, which sets Dorothy's dream in motion, to our journey through **the Quantum & the Dream**, is the statue of Shiva standing on the grounds of CERN, its pose reflecting the cosmic, spiral dance of transformation.

TOTO: A KEY UNCONSCIOUS ALLY

In one of the most frightening scenes from the movie—one I remember rattling my six-year-old-nerves as I watched its TV debut in 1956—is when the wicked witch snarls at Dorothy, "I'll get you, my pretty. And your little dog too," finishing with an ominous, cackling laugh.

While the spiraling tornado initiates Dorothy's call into the dream world of transformation, her little dog, Toto, is actually the most influential agent of change. In the bleak, gray, parched land of Dorothy's Kansas home, Toto is the only bright light in her life. As Baum writes in the opening scene from the original book, "It was Toto that made Dorothy laugh and saved her from growing as gray as her other surroundings."

In the movie version, the wicked neighbor's determination to have Toto destroyed for biting her is Dorothy's motivation to run away from home with him, putting them both in the path of the spiraling tornado which transports them into the dreamscape of Oz.

It is Toto's escape from the Wicked Witch's castle, and his bringing back the Scarecrow, Tin Man, and Lion to knock down the locked door of the death chamber where Dorothy is imprisoned, that saves her life.

It is Toto's "sniffing out" the fake wizard by pulling the curtain aside, exposing the levers used to project a giant floating head onto a screen and a microphone to generate a loud, booming voice, which reveals an important truth the dream is offering. In fact, in this scene, Toto is symbolizing the power of the dream itself, which comes in the service of helping us see through the smoke and mirrors of our limited, egoistic illusions.

It is Toto's jumping out of the hot air balloon basket near the end of the narrative that leads to the most significant insight Dorothy's dream has to offer: Although the wizard agrees to get her back to her home in Kansas with his hot air balloon, when Toto leaps out to chase a cat and Dorothy goes to retrieve him, the balloon takes off without her (like his balloon, the wizard was always full of hot air).

This prompts the reappearance of the Good Witch who reveals the power to return "home" is not a material, physical force, but one generated from within, and is always available if we can recognize it.

Common in dream interpretation is viewing the appearance of a dog as symbol of the dreamer's inner support (man's best friend) and inner guide. "Sniffing" out the fake wizard can be seen as a metaphor for our right-hemisphere oriented intuitive sense which can see beyond the limitations of our physical senses.

Dorothy's overtly emotional bond with her beloved dog adds to the strong feminine dynamic of her Hero's Journey.

THE MELTING

The scene from *The Wizard of Oz* many find most memorable is when Dorothy, in response to the Wicked Witch's setting the Scarecrow on fire, grabs a nearby pail of water and inadvertently splashes the Witch, who starts literally melting away.

Unlike the traditional, long-standing male Hero's Journey, Dorothy doesn't use physical strength, forceful ego, rational planning, or pure willpower to cut down the evil witch: instead, it's her spontaneous, instinctive, inner compassion and emotional love for her companion that motivates her throwing the water. She had no intent to kill.

In fact, after the witch has completely melted and her uniformed, spear-carrying guardsmen approach, she actually apologizes to them, saying, "I didn't mean to kill her." (Can we expect this reaction from Achilles, Hercules, or Rambo?)

To her surprise the guards (all male) bow down to her, shouting "Hail, Dorothy."

In some spiritual teachings, particularly alchemy, "melting" symbolizes the breaking down of entrenched parts of the ego which no longer serve, opening one up to deeper, more core aspects of one's being (in light of **the Quantum & the Dream** synchronicity, this can be projected as the melting of the materialist paradigm).

L. Frank Baum's lifelong interest in Asian spiritual teachings likely introduced him to water as one of Nature's most significant elements of transformation. In dream interpretation, water often symbolizes the dreamer's emotional state. One of the seminal resources we witnessed on our trip so far, which Niels Bohr, the father of quantum theory, cited as a major influence, is the *Tao te Ching*, which often celebrates water as a key transformative power of Nature, noting that its resourcefulness comes from softness and ability to yield:

> Nothing in the world
> is as soft and yielding as water.
> Yet for dissolving the hard and inflexible,
> nothing can surpass it.
> The soft overcomes the hard;

the gentle overcomes the rigid.
Everyone knows this is true,
but few can put it into practice.[135]

The inherent power of softness ascribed to water spiritually and alchemically can also be traced forward to an underlying transformation taking place here in the twenty-first century, right before our eyes on our digital screens:

> One of the main effects of digitization is to make 'liquid' everything that is solid. Anything that can be digitized can be translated into anything else that can be digitized ... as fewer and fewer material conditions present resistance to the storage and delivery of information, the fluidity of digital data brings it as close to the condition of thought as anything we experience in our own minds.[136]

Dorothy's plunge into the uncertain, paradoxical, entangled web of the dream world shifts the very nature of the typical Hero's Journey to the feminine Yin energy of Nature which provides balance to the entrenched Yang energy of the materialist paradigm. Dorothy as archetype connects to the most ground-breaking, influential environmental book ever published, *Silent Spring,* by female conservationist Rachel Carson in 1962, who was willing to go up against huge, powerful, corporate interests polluting the Earth's vital water resources. Dorothy's feisty, determined archetype also connects to the women's movement of the 1960s, and continues forward to more recent young eco-warriors such as Julia Butterfly Hill and Greta Thunberg.

Perhaps it's a coincidence that Dorothy's transformational dream trip appeared just months after the Freud's influential book on dreams. Then again, perhaps it helps solidify the power of synchronicity, that "falling together in time," as a non-physical, meaningful connection between the human psyche and the external world.

As our trip continues, we'll see how Dorothy's indomitable strength out of vulnerability, emotional bonding, and collaborative instincts informs key insights from two brilliant contemporary women who "pull the curtain" on the materialist paradigm, helping us recognize how we fit into the deeper patterns of Nature and can help grow the seeds of the New Renaissance.

4

PROPOSAL *for a* TWENTY-FIRST CENTURY NEW RENAISSANCE FORMULA

If science did not deny consciousness and
intelligence to Nature, things would look very
different.
—ELISABET SAHTOURIS

For the first time in evolution here on Earth, there are three powerful forms of intelligence operating on the planet: Human Intelligence, Computer Intelligence, and by far the most influential, Nature's Intelligence. To grow the seeds of the New Renaissance will require a new philosophical vision which integrates these three more effectively.

$$NI \left(\begin{array}{c} HI \\ + \\ AI \end{array} \right) = New\,Renaissance$$

The graphic above is an original I had designed as a proposed "view" of how the seeds of the New Renaissance have been planted.

NI stands for Nature's Intelligence, HI for Human Intelligence, AI for Computer Intelligence.

Within the parenthesis, human intelligence plus computer intelligence represents the growing interactions between the two as humanity approached the New Millennium and crossed over into the twenty-first century.

What's missing from the materialist paradigm, but present here, is the interaction between human and computer intelligence occurring WITHIN the context of Nature's Intelligence. We were born out of Nature's evolutionary pattern and AI was born out of the human mind. All three are interacting, creating deep patterns never before experienced on this planet.

The materialist paradigm views Nature as something we humans are separate from and can manipulate for our own material desires. How's that working out?

A new storyline, emerging from both new evidence from empirical science and the imaginative vision of a growing number of right-hemisphere thinkers, is what we now explore.

To start, the continuing myopia of the materialist view of Nature comes in large part from a misreading of one of the most brilliant and influential human observers in history, Charles Darwin. The phrase most closely identified with Darwin, stamped repeatedly into our brains during our school years is: "Survival of the fittest."

First, this description of evolution is not Darwin's. It appears nowhere in Darwin's classic *The Origin of Species.* The phrase came from a contemporary of his, the sociologist Herbert Spencer, after reading Darwin's book. Spencer's description of Darwin's theory as "survival of the fittest" came from Spencer's militaristic take on life in general: "Thus, by survival of the fittest, the militant type of society becomes characterized by profound confidence in the governing power, joined with a loyalty causing submission to it in all matters whatever."[137]

While Darwin did include this phrase nine years later in a subsequent book, he made it clear he did not see the description "fittest" in the context of militant, physical dominance and "submission," but rather the ability to adapt to external circumstances. In his essay, "Darwin's Untimely Burial," one of the most critically acclaimed contemporary evolutionary scientists, Stephen Jay Gould, made an important point: that by "fittest," Darwin didn't mean an individual in superior physical shape. Rather, he meant one better adapted for its environment—more like a puzzle piece than an athlete.

Even Darwin's softening the context of "survival of the fittest" from brute force to solving a puzzle, does not sufficiently open our eyes to the more expansive, "bigger picture" of evolution now provided by some brilliant post-Darwinian observers. Before moving on from Darwin, however, it's important to note that, just as Jung didn't prove Freud wrong, only limited in his view of the unconscious, the expansive views of our place in Nature we are about to explore do not prove Darwin's insights wrong, just limited.

As a useful philosophical overview going forward, we'll see that while the competitive, predator/prey, win/lose dynamic has a role in evolution, there's a deeper, more powerful and influential force operating in Nature known as "symbiogenesis."

This term is formed from the Greek word "sym," meaning "together" and "bios," meaning "life," as symbiogenesis refers to the development over time of mutually beneficial relationships between different organisms, eventually leading to the creation of new, more advanced, complex organisms, or in some cases, a whole new ecological system.

As succinctly put in her book, *Microcosmos: Four Billion Years of Microbial Evolution*, microbiologist/philosopher Lynn Margulis, whom we are about to meet, writes: "Life did not take over the world by combat, but by networking." This insight from Lynne Margulis is well worth contemplating.

Seeing Nature as an object to be used reflects our disconnection from the intricate web of life that sustains us. Before we can get a much-needed, clearer picture of the best ways for human and computer intelligence to interact (HI + AI), the insights offered just ahead can get us back in sync with how we fit into the more powerful and influential patterns of Nature's Intelligence and recognize that whatever we mean by "consciousness" starts here.

On the next leg of our trip, we'll check out the research of two prominent women biologists who express the Dorothy Gale archetype of valiant independence and empathy, pointing us towards philosophical and spiritual gateways to the New Renaissance.

5

NATURE'S INTELLIGENCE

Flashback: I'm six years old, walking to my next-door neighbor to play with a friend. Looking down at the sidewalk, I notice a thin blade of grass growing up in the middle of the cement sidewalk. My brain is confused. How is this possible? How can such a thin, weak blade of grass push its way through concrete?

Flash forward 67 years: searching the Web on my personal computer, I tap into the awesome new chatbot called Chat-GPT, where I type in the question: "How does a thin, physically insubstantial blade of grass manage to grow through a thick cement sidewalk?"

Five seconds later I get a full answer, including information about phototropism, the inherent drive of a plant to seek sunlight, and the microscopic weakness

*in parts of concrete which, while invisible to the
naked eye, creates small pores and air pockets. I take a
moment to contemplate the incredible drive that must
be in that blade of grass to push all the way through a
tight pore in cement in order to reach out to the sun.*

LYNN MARGULIS: "SCIENCE'S UNRULY EARTH MOTHER"

If you're looking for a brilliant, feisty, fearless scientist/philosopher
with spiritual sensitivity to investigate the core of Nature's Intelligence,
Lynn Margulis is a great place to start.

As a female biologist rising in the scientific ranks of a male-
dominated field in the 1960s, her original thesis about the origin of
symbiotic (mutually cooperative) life was rejected by fifteen journals.
But Margulis, like that single blade of grass pushing its way up
through the cement sidewalk, eventually rose to such prominence
that historian Jan Sapp commented, "Lynn Margulis's name is as
synonymous with symbiosis as Charles Darwin's is with evolution."

Margulis, who died in 2011, was a scientific rebel who achieved
many honors in her field, including the National Medal of Science.

She revealed the limitation of Darwin's theory of random
mutation, providing empirical evidence that while there is clearly
intense competition and natural selection operating in evolution,
advanced forms of life emerge out of an even more powerful,
underlying dynamic of symbiogenesis, the process of living together
in a cooperative dynamic where each organism benefits.

What's intriguing about Margulis' early scientific exploration
is that she found this trend of symbiogenesis in simple microbes,
suggesting that complex cellular organisms (such as us, along with all
plants and animals) originated from the merging of different types of
cells living together in a relationship based primarily on cooperation

and collaboration, superseding the "survival of the fittest, winner take all" paradigm so dominant in cultural consciousness.

QUICK THOUGHT EXPERIMENT:

For about 30 seconds, take an inventory of the inner workings of your physical body.

Chances are, virtually all of us doing this would identify major organs such as our heart, intestines, lungs, brain, etc. After all, where would be without these crucial organs?

Compare your internal body inventory with this from evolutionary biologist Lynn Margulis:

> More bacteria inhabit your intestinal tract right now than the number of people who lived on Earth in the last million years. Your body contains a greater number of bacterial than human cells. Some bacteria with tiny magnets in their bodies orient and swim north more accurately than fish. . . . The effects we recognize as sensitivity to light, sense of touch, hearing, smell, and indeed our senses in general evolved from a property properly called 'bacterial consciousness'. . . . Our ultimate ancestors, yours and mine, descended from this group of beings.[138]

As **the Quantum & the Dream** synchronicity induced the parallel descent into the depths of both the unconscious mind and the subatomic realm, Margulis descended into the microcosmic world of bacteria and other ancient single-celled organisms to expand our vision of Nature's Intelligence.

While crediting Darwin for seeing that human evolution was part of a much larger, more complex process than previously believed, Margulis followed her intuitive inclination to study primitive, ancient precellular bacteria. She writes, "In the merger process among bacteria community members, relationships changed: aggression gave way to truce, accommodation followed cannibalism and predation, and cohabitation succeeded in some with great perseverance through the ages."[139]

What brought floods of criticism to her early work was her claim that at the microbial level bacteria act in ways that can accurately be considered "conscious."

Pointing out that "consciousness is 'awareness of the world around one,'" she offers that "Evidence for bacterial awareness abounds in the scientific literature. Many bacteria glide toward oxygen gas and away from sulfide or swim to edible sugars and away from strong acids or dangerously high salt solutions."

(Sounds like bacteria are more conscious than millions of Americans rushing towards heart-damaging heaps of over-salted, overly-processed foods laden with artificial coloring and trans fats.)

Following up on Margulis' insights into the origins of how we got here: "Indeed, we have much to learn from bacteria—including how to prepare soil for plants, recycle nitrogen, and conserve water. . . . Bacteria invented photosynthesis and swimming, evolving prior to any animals or plants."[140]

It's worth pausing here to reflect on Margulis' statement regarding the ability of ancient bacteria to figure out photosynthesis. Photosynthesis could well be the most influential process for creating advanced life on this planet. It was through this process that bacteria developed the ability to absorb energy from the sun, then combine it with water and carbon dioxide to produce nutrients that greatly enhanced their ability to grow and multiply. Through photosynthesis, bacteria learned to use the carbon dioxide they required and then release oxygen, which was toxic to them.

As bacteria proliferated over millions of years, constantly releasing more and more oxygen into the air, they were the organisms responsible for changing the chemistry of the Earth's atmosphere with enough oxygen to support the evolution of plants, animals, and eventually us. In fact, renowned ethnobotanist Dennis McKenna calls photosynthesis "the greatest discovery of all time."

Thanks to those ancient bacteria, today all plant and animal life, including us humans, reap the benefit of a massive, perfectly operating symbiosis: plants take in carbon dioxide and release oxygen that are waste products to them as we take in our much-needed oxygen and release carbon dioxide that are waste products to us, in a perfect, mutually beneficial exchange.

While there are clearly plenty of "wars" among bacteria competing for food and physical advantages within each of our immune systems, as within all species of life, symbiogenesis, the mutually supporting, win/win dynamic is more pervasive and influential throughout Nature for creating more complex-intelligence life forms over time.

Margulis adds to bacteria's resume of accomplishment: "Bacteria are exemplary genetic engineers: splices and dicers and mergers of genome par excellence. We people just borrow their native skills."

While humans have evolved into the most intelligent species, we are still totally dependent on the trillions of bacteria residing in our bodies. Margulis notes that "Like nearly all other animals, we mammals harbor in our intestines an assortment of specific bacteria that help us digest our food. . . . Without these hitchhikers to help digest fiber and produce vitamins . . . we weaken and even die."[141]

What virtually all of us tend to regard as lowly, annoying creatures, precellular brainless bacteria are in fact an illuminating example of how we reside within Nature's Intelligence and Nature's Intelligence resides within us as we spiral towards Evolution's next creative leap.

We've refined processes like photosynthesis and genetic engineering, but it's the bacteria both inside of us and around us that adapted them first and from whom we learned how to create many of our best technologies.

Lynn Margulis, in her assertive, "suffer no fools" style, writes, "Those who hate and want to kill bacteria indulge in self-hatred. Not only are bacteria our ancestors, but also, if I am correct, as the evolutionary antecedent of the nervous system, they invented consciousness. . . . Our cultural prejudices, our haughty depreciation, prevent access to their ancient wisdom and the salutary effects of dialogue."[142]

Anyone who pays attention to basic nutrition or has consulted with a clinical nutritionist knows that, while the invention of antibiotics has extended the life span of humans significantly by killing illness-threatening bacteria, massive overuse has killed off the much larger population of good bacteria which are essential partners in our overall health and well-being. In fact, our health is totally dependent on a "healthy gut," in which over 100 trillion microbes create a balanced environment for digesting food, metabolizing nutrients and reducing inflammation.

Willing to push through the hard resistance to her early work, Lynn Margulis' discovery of consciousness and intelligence at the microbial level helps us get a better feel for how dependent we are on Nature's inherent drive to create more complex and novel forms. She certainly earned the epithet "Science's unruly Earth Mother."

Which provides a good transition to another of her influential roles.

In 1971, at the request of chemist James Lovelock, who had been writing essays on a controversial, holistic view of how life on Earth operates, Margulis collaborated with him to flesh out what Lovelock called the Gaia Hypothesis (Gaia being the ancient Greek mythological name for the personification of Earth). Its basic premise, which has generated both significant praise and angry dismissal from the scientific community, suggests that understanding the planet Earth as a living, self-regulating, biological organism is more scientifically and philosophically accurate than the materialist view of Earth as an inanimate object best understood by breaking it down into its separate parts, such as the atmosphere, soil, oceans, etc.

The Gaia theory is quite technical, requiring an advanced knowledge of chemistry and biology which I certainly don't have. In terms of **the Quantum & the Dream** synchronicity, we can see how Lovelock and Margolis' empirical research and philosophical views resonate with the first shift we observed on our trip: from the left hemisphere of the brain to the right hemisphere. Understanding the entire Earth as a living, interconnecting organism rather than material series of separate parts reflects the kind of right-hemisphere vision of the early quantum theorists we looked at earlier: "Quantum theory thus reveals a basic oneness of the universe. It shows that we cannot decompose the world into independently existing smallest units. As we penetrate into matter, nature does not show us any isolated 'basic building blocks,' but rather appears as a web of relations between the various parts of the whole."[143]

The Gaia Hypothesis further connects to **the Quantum & the Dream** synchronicity in another interesting way. Just as Freud initially, then Jung subsequently, saw ancient mythological themes as reflections of dynamic activity in the human unconscious, Lovelock, supported by Margulis, adopted the mythological name Gaia to resonate with the "big picture" perspective of Earth as a self-organizing, dynamic system.

Helping us to see our human position not as superior and apart from Nature, but as a living part of its deeper intelligence, is a crucial shift in perception if we are to seed the New Renaissance. Margulis reminds us that "The trip from greedy gluttony, from instant satisfaction, to long-term mutualism has been made many times in the microcosm. . . . Indeed, it does not even take foresight or intelligence to make it: the brutal destroyers destroy themselves, while those who interact more successfully inherit the living world."[144]

Lynn Margulis was a fiercely independent thinker and assertive personality, a powerful personification of the "Dorothy Gale archetype" which emerged, along with Freud's book on dreams and Planck's discovery of the quantum in the first year of the modern age, 1900. Like Dorothy, Margolis balanced her bold self-reliance with enthusiastic collaboration, as demonstrated with her extensive work with James Lovelock on Gaia Theory and with the books she collaborated on with her son, Dorion Sagan.

Before moving on from her contribution to our understanding of Nature's Intelligence, here's another synchronicity which occurred as I was starting to read Margolis' book of essays *Dazzle Gradually: Reflections on the Nature of Nature*.

The first essay is a short autobiographical sketch, "Red Shoe Conundrum," which immediately flashed me back to Dorothy's iconic <u>red</u> ruby slippers, a key motif in *The Wizard of Oz*.

Margulis describes how, growing up in the late 1940's, she had an independent streak: "Unlike many friends, neither as an adolescent nor as a young adult did I wait for 'my prince to come.'"

During this period, she took weekly excursions to ballet classes and writes that the one film that had the greatest impact on her was *The Red Shoes,* the passionate and tragic story of a talented ballerina based on a Hans Christian Anderson fairy tale, just as Dorothy's adventure takes place in a fairy tale setting.

In the movie's climatic scene, the enchanted red shoes take on a life of their own, forcing the ballerina to literally dance to her death, leaping off a ledge in front of an oncoming train. Margulis writes,

> The talent of this beautiful ballerina in the prima
> donna role was exhilarating, as was her true love for
> her sexy, handsome beau. I remember feeling anger
> at the melodrama of that movie, however. I thought
> the dichotomy of her life that led to her self-instigated
> fate utterly ridiculous. Why did there have to be
> "necessity to choose" between devotion to a man and

devotion to a career? What generated the psychic dissonance that distracted her to destruction?[145]

Margulis continues,

If the star had been male, he would not have been driven to choose. . . . Instead, under relentless pressure to be the perfect dancer whose shoes run away with her, the ballerina yields to the dance master's demands that she remain in the spotlight of his world while her lover demands that she marry him and have a family. . . . Why hadn't she simply married her lover, borne her children, and continued dancing?[146]

Which is exactly what Margulis ended up doing. At a young age she married the man who would become a world-famous astronomer, Carl Sagan (whom she eventually left, as she did two subsequent husbands), raised four children, and "never abandoned science for even a single day in over forty-five years."

One could say she psychologically and philosophically traded in those self-destructive "red shoes" of the tragic ballerina for the dream-enhancing ruby red slippers which gave Dorothy solid footing on her heroic journey of self-discovery. The color red has long held the symbolic power of the life force (blood circulating through the body), often appearing in myths, fairy tales, and dreams to reflect fiery emotion, sometimes a fire which rages uncontrollably, sometimes the potential for psychological/spiritual transformation. Carl Jung titled his remarkable journal reflecting his dedication to exploring the depths of his unconscious, *The Red Book*.

Like Dorothy, Lynn Margulis fearlessly pursued what she intuited to be her soul's path.

ELISABET SAHTOURIS: "WE ARE CAPABLE OF REGAINING OUR REVERENCE FOR LIFE."

Evolutionary biologist Dr. Elisabet Sahtouris has been a global influencer, helping to expand the emerging post-Darwinian "bigger picture" of how we got here, where we are and where we may be headed.

Sahtouris is known for her research and advocacy for two of the key insights introduced by Lynn Margulis: the role of symbiosis in the flow of Evolution and the Gaia hypothesis, exploring how different organisms interact within ecosystems, creating a delicate balance that supports life through an intricate web of relationships, shining the light of conscious awareness to Nature's Intelligence.

Sahtouris has worked tirelessly outside the classroom, giving talks around the world, advising ethical markets, taking leadership roles as a founding member of Worldwide Indigenous Science Network and Rising Women, Rising World, and advisor to ecological groups such as the Earth Restoration Corps.

In an extended interview she both celebrates and adds to the "bigger picture" provided by Lynn Margulis:

> We have a hundred trillion cells in our bodies and each one of them has thirty thousand recycling centers renewing our proteins. They're so hi-tech that they can take in a protein, disassemble it, build a new protein (perhaps an entirely different kind) and issue the new protein. That's as if we could stick trees into a chipper machine and get a live tree out the other side. Very hi-tech![147]

As with Margulis, Sahtouris sees through the limitations and harm of the materialist paradigm and supports the provocative view that by peering more deeply into the patterns of Evolution, we can detect in Nature a conscious intelligence.

What I see happening... is recognition of consciousness/intelligence at the molecular level creeping into microbiology just as the notion of Earth as alive (conscious and intelligent in its own right) is sneaking up on systems theorists in microbiology. In my own opinion, the universe, like any living system, operates on conscious micro and macro levels at once. This is, of course, still pretty heretical.[148]

Echoing the call for a shift from the territorial-oriented rationality of the left hemisphere to the more expansive, visionary right hemisphere, Sahtouris offers an intriguing thought experiment which furthers the perception that, while we are conditioned to view ourselves as separate from Nature, our individual bodies are a microcosm of a living planetary organism:

If you could see a picture of Earth in a few hours, as it's been from the beginning, you'd have no doubt that this is a living entity constantly changing and recreating itself, and evolving evermore complexities. Three-quarters of its life was devoted just to microbial life and then the big multi-celled creatures came in. The Earth itself is like a giant cell. Even Redwoods have just a thin skin of what we call biological life on its surface. The rest of it isn't alive by our definition. And yet we think of the whole tree as alive. So, the planet with its thin skin of biological activity also seems to be a self-creating kind of cell.[149]

(Note: The magnificent photo of the entire Earth, known as "The Blue Marble," taken by the crew of the 1972 Apollo moon mission is one of the most widely distributed and impactful photos ever produced. Yet, it was a still photo. In the "thought experiment"

Sahtouris encourages, we see "The Blue Marble" as Gaia, a living, breathing organism emerging out of all of her interacting parts, including us.)

Sahtouris' thought experiment also offers a deep recognition of how evolution transcends conflict and competition to develop a deeper pattern of mutually beneficial synergy:

> Humanity, as it diversified and had more and more people, created more and more conflict. Exactly as the early Earth differentiated into bacteria and then they developed different lifestyles and they became competitive. They invented technologies in order to carry out their hostilities. They created enormous problems including global hunger and global pollution. And had to solve those eventually by negotiating differences, moving on around the cycle, and working out cooperative schemes that ultimately lead the ancient bacteria that ruled for the first half of Earth's life to form a new kind of cell as a community of different lifestyle bacteria working together. That's the nucleated cell that we're made of, that all these trees are made of, all the beings in the waters are made of. Everything we see around us is made of this wonderful big cooperative cell.[150]

The reference to "all these trees" to reflect Nature's Intelligence connects back to the suggested thought experiment at the beginning of our trip through **the Quantum & the Dream** synchronicity, to imagine we can peer beneath the ground to the root system of an oak tree, suggesting an analogy between the tree's ability to absorb nutrients from the soil with the ability of our right-hemisphere to absorb creative insights through the thought experiments, reveries, intuitive flashes, and dream patterns percolating up from our unconscious.

Turns out that "all these trees" referred to by Margulis had already developed a complex communication system among their roots well before we showed up on the planet. In 2018 a German forester, Peter Wohlleben, discovered yet another key example of Nature's Intelligence which had eluded scientists. As reported in *Smithsonian Magazine,*

> A revolution has been taking place in the scientific understanding of trees and Wohlleben is the first writer to convey its amazements to a general audience. The latest scientific studies, conducted at well-respected universities in Germany and around the world, confirm what he has long suspected from close observation in this forest: Trees are far more alert, social, sophisticated, even intelligent than we thought.[151]

The article goes on to confirm a significant understanding of Evolution:

> Since Darwin, we have generally thought of trees as striving, disconnected loners, competing for water, nutrients and sunlight, with the winners shading out the losers and sucking them dry. The timber industry in particular sees forests as wood-producing systems and battlegrounds for survival of the fittest. There is now a substantial body of scientific evidence that refutes that idea. It shows instead that trees of the same species are communal. . . . These soaring

columns of living wood draw the eye upward to their outspreading crowns, but the real action is taking place underground, just a few inches below our feet. "Some are calling it the **Wood-Wide Web**," says Wohlleben.[152]

I'm fascinated by the fact that I learned about the "Wood-Wide Web" awareness of underground tree roots by accessing the World Wide Web on my computer's digital screen. This Wood-Wide Web/ World Wide Web connection offers an interesting analogy to the formula proposed at the beginning of this segment of our trip:

$$NI \left(\frac{HI}{+ AI} \right) = New\ Renaissance$$

If, as Margulis and Sahtouris reveal, ancient microbes could develop photosynthesis and the nucleated cell, then can't we, living in the era of the World Wide Web—through which billions of human brains are sending research data, ideas, insights, dreams, and imaginative visions instantaneously around the planet—help induce the New Renaissance?

Just as Lynn Margulis balanced her fiery, independent streak with a talent for collaboration, so Elisabet Sahtouris combines her drive as an independent woman researcher unafraid to challenge entrenched scientific and philosophical paradigms with a talent for collaborating with professionals from diverse backgrounds. A good example is the book, *Biology Revisited*, an extended dialogue between her and the late Willis Harmon who was president of one of the most forward-thinking organizations integrating science, philosophy, and spirituality, the Institute for Noetic Sciences.

In this extended conversation, Sahtouris supports a key theme we've been absorbing throughout our trip—the eroding of the materialist paradigm which insists on viewing the world as a series of objects we are separated from and can manipulate for our own, selfish ends: "We have been taught that our planet and all its creatures were not conscious or intelligent before humans appeared. We have been taught to look at nature as though it really were assembled from particles into part and from parts into wholes, like machinery."[153]

She follows this up with an insight that is crucial if humanity is going to fully seed the new renaissance:

> If science did not deny consciousness and intelligence in nature, things would look very different. We could no longer see nature as an array of resources for human use; we could no longer see ourselves as the only intelligent beings. We humans would not be in full control and our technological solutions to problems would often come into question.[154]

It's not only science that needs to move past the eroding, antiquated, myopic materialist paradigm, but philosophy, psychology, and the nineteenth century factory-model educational system still in operation today. We will see evidence of this trend further ahead.

In tune with **the Quantum & the Dream** synchronicity, Sahtouris makes it clear that to understand consciousness, we need to have a deep understanding of the unconscious mind, and she particularly underscores the value of the collective unconscious, which she and her colleague William Harmon describe as being "analogous to the membrane delimiting a cell, which permits a constant and active interchange with the whole body to which the cell belongs; it allows a sort of 'psychological osmosis' with other human beings and with the general psychic environment."[155]

The combined research and insights of Elizabeth Sartorius and Lynn Margulis provide a profound recognition of the symbiotic intelligence inherent in nature, within which life forms transcend random mutation and predator/prey battles to collaborate from within a self-organizing, inter-connective core.

This intricate web of relationships showcases nature's intelligence in maintaining equilibrium, adapting to changes, creating novel forms of biodiversity. Philosophically, the Gaia Hypothesis suggests that we Homo sapiens, having arrived on the evolutionary scene with the immense gift of a complex cerebral cortex featuring a right hemisphere capable of self-conscious insight and imaginative leaps, have a most important, responsible role in what happens on this planet going forward.

In her essay "Prologue to a New Model of a Living Universe" Elisabet Sahtouris makes a passionate case for putting mind before matter: "The best argument we have for the existence of a "real" vast universe is the limitlessness of human conscious awareness, whether it is focused inward or outward. Every scientific or spiritual discovery can be contained within its expansive capacity."[156]

6

BRIEF INTERLUDE:
WHO'S TAKING FLIGHT?

Austrian Johannes Fritz is known as a "maverick biologist." Like Lynn Margulis and Elisabet Sahtouris, he has a deeply-felt sense of Gaian awareness. From a young age he had the desire to reintroduce birds who only existed in captivity back into Nature. He fell in love with a breed called the northern bald ibis, the size of a goose with an extremely long beak, and considered a very unattractive bird. Fritz was one of the few humans willing to take responsibility for raising and caring for them.

The northern bald ibis had been hunted and eaten into virtual extinction. The only ones remaining had to be raised in captivity. To release the birds he was caring for back into the wild, Fritz had to first teach them how to migrate south in the Fall, as they would never survive the cold Austrian winter on their own. He learned how to fly an ultralight aircraft, then remodeled it to fly as slowly as twenty-five miles per hour, the maximum speed at which the northern bald ibis can fly.

The altered ultralight, now even lighter, was basically a small cockpit with a propeller and canopy overhead.

After training the birds to follow him into the air (he was imprinted in their brains as being their mother), Fritz led them south across the Alps to an area in Italy warm enough in winter for them to prosper. From then on, the birds were able to migrate on their own. Fritz extended his personal bald ibis teaching tour to over 250 additional pupils over the years.

Then, the growing effects of climate change made the traditional migration route impossible. The warmer Fall season on the Austria-

Germany border resulted in the birds waiting a month longer to start the trip south for winter. Following one of the flocks in his ultralight, Fritz could see that the mountainous Alps region was now too icy for his beloved birds. Frosted ground was making it difficult to find worms, and snow was covering the ibis' feathers. They would perish. Fritz and a small team he assembled rounded up the struggling birds, crated them, and brought them back.

He studied weather maps and is devising a new route south the birds will be able to survive as the European continent climate continues to get warmer. The proposed new route will be three times longer, well over two thousand miles, heading west to France, then south along the Mediterranean coast to Spain's southern peninsula. The new trip in the ultralight leading the ibis flocks will likely take at least six weeks.

The new teaching route will be risky. Fritz has already survived a crash in a cornfield, and the weather along the new, extended route will likely have periods of rough weather for such a fragile, ultralight aircraft. But he is undeterred. His two sons have also learned to fly the remodeled ultralights and will help lead the ibis flocks on their new migration route.

When interviewed about the upcoming venture, Fritz said, "It's not so much a job, but my life's purpose."[157]

This account is a fabulous teaching tale on its own. It's also intriguing to consider it in a "bigger picture" context, as well: Fritz's amazing love for Nature became better known through an article posted on the World Wide Web, where one can also read about the discovery of Nature's vast communications system of underground tree roots cleverly referred to as "the wood-wide web." Both of these "webs" are accessible to the human brain, previously described on our trip by physicist/philosopher Fritjof Capra as *"intricate patterns of intertwined webs, networks nesting within larger networks."*[158]

Continuing to expand this vision, the World Wide Web, Wood-Wide Web and our human brain's "patterns of intertwined webs" are all operating within the Gaian web of Nature's Intelligence.

7

HI & AI
(HUMAN & COMPUTER INTELLIGENCE)

Success in Circuit lies
Too bright for our infirm Delight
The Truth's superb surprise
—EMILY DICKINSON

As Depth Psychology has shown, we humans can't help but unconsciously project the inherent fears and anxieties in our memory banks onto the outside world.

While there are clearly significant risks at play with the continuously increasing influence of AI on every aspect of our lives, one of the key underlying causes of AI anxiety comes from ancient and contemporary storylines about human-made artifacts which run amok or go awry. Some examples:

- From ancient Greek myth, Talos, a giant automaton made from bronze, created to protect the ancient Green Isle of Crete from invading ships, but who became bewitched by the sorceress Medea and was destroyed.

- The Golem, an ancient Jewish folklore about an animated being made from mud or clay by humans which then becomes uncontrollable, sometimes a villain, sometimes a victim.

- The Terminator, a cyborg assassin sent into the future to kill a woman whose unborn son will one day save humanity from extinction (the director James Cameron, said the character came to him in a "fever dream").

- HAL, the infamous super computer-turned-murderer, in *2001: A Space Odyssey* (for our final thought experiment we'll descend into HAL's brain).

What makes computer intelligence unique in human evolution is its incalculable effect on our psyches, as it is not just a material artifact, but a new form of intelligence we're uncertain can be controlled as it continues to expand at an unimaginable speed.

It's mind blowing to try to comprehend that during the second half of the twentieth century vacuum tubes morphed into transistors, transistors into semiconductors, semiconductors into advanced integrated circuits, which generate the speed at which computers can access and analyze information from twenty times per second to (fasten your seat belts) ... over a billion times per second.

And quantum computers, expected to become mainstream by 2030, which are predicted to be well over a million times faster.

How do we wrap our minds around this?

Another part of the unconscious discomfort with computer intelligence comes from its being referred to as "artificial intelligence." To be artificial has the connotation of "false," "imitative," "stilted," "deceptive," "fake."

John McCarthy, who coined the term "artificial intelligence" in 1956, later regretted the term when he saw how his original meaning was totally misinterpreted. His intention was to make reference to the distinction between the human biological brain and the potential mechanical intelligence humans could make out of hardware.

McCarthy chose the word "artificial" as the adjective form of the noun "artifact." An artifact refers to an object, a tool or a craft designed by human ingenuity. McCarthy, often referred to as "the

Father of Artificial Intelligence," meant by his enduring term, not fake or inauthentic intelligence, but intelligence crafted by human ingenuity.

Given the enormous consequences, creative and destructive (can we hear the echoes of Shiva's dance?) and the unprecedented speed of change we have to adjust to due to AI, the rising level of anxiety coursing through contemporary cultures is not surprising.

Much of the current anxiety over AI is the threat of it taking over millions of paying jobs people need to support themselves and their families. There is no question AI will be causing huge dislocations in the economic markets. But as we saw with the shift taking place from the printed page to the digital screen, virtually every major new technology throughout human history has been disruptive to the status quo and, after a period of adjustment, has proved to be a boon for humankind.

One significant example: In the year 1800, 90 percent of all Americans lived and worked on farms. By the year 1900, with the Industrial Revolution well under way, only 40 percent lived and worked on farms. What happened to those millions of Americans forced out of work by the new steam-powered machinery disrupting the status quo? Many took jobs created by new industry, others located to growing cities to take advantage of work offerings there. The educational system expanded to train people who needed new skills. By the year 2000, the percentage of Americans living and working on farms was less than 2 percent, yet millions of good new jobs emerged that hadn't even been thought of one hundred years before. (One sign of the New Renaissance will be the significant increase of "back to the land," small farms starting up.)

This is a scenario worth our attention: America went from 90 percent of people living and working on farms to less than 2 percent in only about 10 generations. Many people "lost" factory and assembly-line jobs that were mind-numbing and physically dangerous ones, moving into new work opportunities. To give just one contemporary example, according to Deloitte, the world's largest service network, although technology has displaced over 800,000 jobs in the UK between 2001 and 2015, it has created approximately 3.5 million new ones.

The deeper level of anxiety over AI, however, is at the psychological/mythological level. The prevalent fear that supercomputers might evolve into formidable predators comes in large part from our projecting onto AI the long, historical record of our species using our superior intelligence to dominate and exploit all other life on the planet for our own needs, subjugating other species and even other human cultures. So, we assume/project that computers will act the way we have and we'll have to compete, or even go to war with AI as it gets more and more intelligent.

According to a May, 2023 Reuters/Ipsos poll, 61 percent of Americans say AI threatens humanity's future.

The age of AI on the horizon with genetic engineering, nano-technology implants, and quantum computers operating a million times faster could radically diminish what we have assumed is our rightful place at the top of the evolutionary pecking order. AI, the "artifact" we created, could get totally out of our control. How much of this anxiety is real, how much unconscious projection?

In depth psychology, projection is a most important phenomenon to understand. One of the most spiritually-oriented acts we can perform is to recognize and "own our projections," that is, be more aware that it is likely taking place, a reminder that most of our negative perceptions come from inside us, generated by memories on the personal level which are unique to each of us and also at the collective unconscious level which has generated the enduring mythological stories and teaching tales over the millennia and which all humans, to some degree, share.

In an interview, Yuval Noah Harari, author of *Sapiens: A History of Humankind*, agrees that a key for the future relationship between human and computer intelligence will take place within the human psyche: "For thousands of years, we have gained the power to control the world outside us but not to control the world inside."[159]

The more we can see through the limitations of the materialist paradigm to the deeper, more accurate picture of the symbiotic pattern at the core of evolution's drive towards novel, more complex, collaborative life, the more conscious we can become of our negative projections—and the more we can maximize the potential benefits of combining human and computer intelligence.

Nature has given us, Homo sapiens, the responsibility of having a hand on the steering wheel of evolution. A lot will depend on our increasing ability to recognize our fearful, anxious "projections."

Following are a quick thought experiment and two powerful examples of mutually beneficial, symbiotic relationships between human and computer intelligence. One is so well known that its historic benefits are often taken for granted. The other, not yet well known, is truly amazing.

"THE ECOSYSTEM OF FREE KNOWLEDGE" THOUGHT EXPERIMENT:

It's the year 2000. A friend calls, his voice filled with excitement about a New Millennium idea. Fascinated by the growing phenomenon called the World Wide Web, this friend raves on about creating a nonprofit online encyclopedia, the largest and most expansive ever, but instead of hiring dozens of researchers and editors, he will count on attracting thousands of unpaid volunteers around the world to post articles. As for editing, while there would be a small group of hired editors to take out any offensive or clearly bogus content, the vast majority of editing would also be an undirected free-for-all, with anybody able to edit anybody else's posted articles. Your friend, getting even more excited, goes on to tell you he expects this volunteer, user-created, digital encyclopedia to gain the status of the esteemed *Encyclopedia*

Britannica, considered the gold standard reference source, which has the budget for hundreds of paid contributors and editors, including over one hundred Nobel Prize winners.

What are the odds your reaction would be something like, "ARE YOU NUTS?"

Wouldn't this virtually unsupervised, unpaid volunteer-generated encyclopedia be a total chaos-filled mess, filled with prejudiced hype and made-up facts?

Anyone can place an article?

Anyone can edit an article?

To make the whole scenario even more delusional, your friend states his goal to have this fantasy research tool help create "a world in which every single person on the planet is given free access to the sum of all human knowledge."

Turns out this lunatic idea, launched in January of 2001, now has over fifty million articles in over three hundred languages, is highly regarded, and two years after launching, if your friend in the thought experiment was Jimmy Wales, he was named by *TIME* magazine as "one of the one hundred most influential people in the world." His and partner Larry Sanger's fantasy, "crowd-sourced" Wikipedia, is now accessed every day by more than forty-six million mobile devices and more than twenty-three million users on desktop computers, and its pages have been edited over one billion times.

As for its accuracy, Wikipedia's science articles were compared to those in the esteemed *Encyclopedia Britannica*, and found to be at relatively the same level. According to a joint pilot study conducted by Oxford University and Epic, an e-learning consultancy, "Wikipedia fared well in this sample against *Encyclopedia Britannica* in terms of accuracy, references and overall judgement, with little differences between the two on style and overall quality score."[160]

Forbes magazine reported, "Perhaps the most interesting finding ... is that the more times an article is revised on *Wikipedia*, the less bias it is likely to show—directly contradicting the theory that ideological groups might self-select over time into increasingly biased camps."[161]

Kevin Kelly, cofounder of *Wired* magazine, and one of the most respected chroniclers of AI, was so impressed that he wrote "Wikipedia has taught me to believe in the impossible more often."[162]

How is this possible? How can such an undirected, chaotic environment where anyone can post an entry (an average of over five hundred articles are added every day) and anyone can edit anyone else's entries (as of this writing edits have been made to over fifty million entries), with only a skeletal staff of paid editors supervising and removing offensive language and clearly bogus content, become the greatest aggregator of knowledge and wisdom ever created?

Kevin Kelly gets to the gist of how the underlying force of Nature's symbiotic, mutual benefit pattern can operate when human intelligence and computer intelligence maximize each other's attributes:

> This list goes on, old impossibilities appearing as new possibilities daily. But why now? What is happening to disrupt the ancient impossible/possible boundary? As far as I can tell, the impossible things happening now are in every case due to the emergence of a new level of organization that did not exist before. These incredible eruptions are the result of large-scale collaboration and massive real-tie social interacting, which in turn are enabled by omnipresent instant connection between billions of people at a planetary scale.[163]

The phrase Kelly chooses, "...instant connection between billions of people at a planetary scale" is echoed by communications expert Derrick de Kerckhove, Director of the McLuhan Program in Culture and Technology:

The pressure of human minds focusing on the same issues and the self-organizing abilities of the network create a potential for a great unity of purpose. All these organic minds can be assisted by digital media which vastly increase their power of synthesis and classification. The significance of Web is not that it is yet another distribution system, but that it is a *distributed* system. The fun and the substance of the Web is in its ability to connect living minds at work in all manner of purposeful configuration. The minds on the Net are connected and they behave like liquid crystal in stable but fluid formations.[164]

We now live in an age where human intelligence, combined with the expanding collaborative platforms generated by AI, are achieving results previously deemed impossible. It's true that a majority of the most popular websites remain under the control of huge corporations seeking more and more profits and control over users. It's true that governments in less developed countries are forcefully controlling Web content. Meanwhile, Wikipedia, on the wave of tens of thousands of passionate unpaid volunteers around the world, posting and editing entries within the stated guidelines of N.P.O.V (neutral point of view), has remained one of the top ten most visited websites in the world.

At a deeper level, this is a beautiful example of Nature's principle of self-organizing intelligence in operation.

In an interview, Jade, one of the Wikipedia volunteers who puts ten to twenty hours a week into editing articles and has over 24,000 edits to her credit, conveyed the credo she shares with so many Wikipedia volunteers dedicated to sharing knowledge: "My calculations in the past are, you know, more than 10 million people read my work in a year, so it's an honor to have people reading all that."[165]

Postscript: Wikipedia supplanted the *Encyclopedia Britannica* which, after publishing every year since 1768, announced in 2012 that it was discontinuing its print publication and is now only available online.

The very act of connecting people's minds on a global scale, with virtually no physical, geographical limit and no central authority, and with an open-ended, self-organizing feedback loop, is an evolutionary game-changer.

De Kerckhove, clearly influenced by his mentor McLuhan, adds a psychological context to the growing planetary connectedness of human and computer intelligence, identifying the positive side of "projection:"

> We are continuously projecting ourselves via all of our senses, and with the electronic extensions of our hands, eyes, ears, and voice, we have acquired the ability to project ourselves far beyond the limits of our bodies and to receive the projections of others as if we could 'wear' them. Our new skin is very sensitive; it is made of the millions of interactions of computers and electronic webs all over the planet. This is a tactile world. Indeed, the world is not 'out there' anymore, it is right here, under the skin of each of us. Now that we are extended beyond the boundaries of our biological being, it follows that eventually our psychological makeup will be modified accordingly.[166]

As previously proposed on this trip, the Web, now connecting over five billion human minds, can be seen as an extension of the collective unconscious being projected back to us through our digital screens. Now that we are, in De Kerckhove's term, "extended beyond the boundaries of our biological being," how are we going to react to this, individually and collectively, going forward?

For those looking for a ray of optimism in the face of the continuing battle of mega corporations and political forces thirsty for even more control and influence over those of using the Web, just three individuals, none of them wealthy or with any access to power, have had a seismic cultural impact on the current age. And they have had this impact by thinking beyond individual wealth and glory, reflecting Nature's evolutionary pattern which, over time, transcends the continuous competition for resources to a deeper core instinct for symbiotic, mutually beneficial collaboration.

We have already identified two of these people: Jimmy Wales and Larry Sanger, seemingly naïve visionaries with the preposterous idea of creating the most expansive resource of free, retrievable knowledge and wisdom ever devised, through the tireless passion of thousands of unpaid, primarily self-organizing volunteers.

The third, whom we also met previously on our trip, Tim Berners-Lee, single-handedly created the World Wide Web in 1989 and, in agreement with his supervisor at the CERN particle accelerator where he was working, chose not to patent and monetize it and, acknowledging it was too important a tool for the benefit of humanity to be "owned" by a few, put it in the public domain.

These three have heeded the calling of journalist Adrienne LaFrance in her essay "The Coming Humanist Renaissance":

> We should know by now that neither the government's understanding of new technologies nor self-regulation by tech behemoths can adequately keep pace with the speed of technological change or Silicon Valley's capacity to seek profit and scale at the expense of societal and democratic health. What defines this next phase of human history must begin with the individual.[167]

The enormous positive influence of Jimmy Wales and Larry

Sanger with Wikipedia, and Tim Berners-Lee with the protocols underpinning the Web itself, is not just the content it creates, but even more importantly, the expansion of a free, open-source digital platform, on which unprecedented levels of mutually beneficial collaboration is expanding.

An even "bigger picture" of the success of Wikipedia and the vision for humanity of Berners-Lee reveals the enormous impact a few individuals can now have in the digital age of AI. Adding emphasis to this recognition, cultural historian William Irwin Thompson observed, "For the first time in human evolution, the individual life is long enough, and the cultural transformation swift enough, that the individual mind is now a constituent player in the global transformation of human culture."[168]

On the next leg of our journey, we will see how another individual with a seemingly impossible vision formed a team which is literally connecting human and computer intelligence with Nature's Intelligence.

It begins with a horse race.

8

THE WINNING TICKET:
"WE AMPLIFY INTELLIGENCE"

We believe the most important database on the
planet is the diverse knowledge, wisdom, and
insights housed within the billions of human
minds distributed around the globe. Every
one of us has over a million gigabytes between
our ears, and we are continuously interacting
with our environment, testing our hypotheses,
reevaluating our assumptions, and updating our
vast data-stores. It's incredible.
—UNANIMOUS AI WEBSITE

When I first discovered Unanimous AI while searching the Web back in 2017, I found it too fantastic to be true. It does check out, and over the past few years, while still largely not on the mainstream radar, is gaining influence with major organizations and companies.

Its motto is: "We amplify intelligence."

They do it by tapping into Nature's Intelligence.

Unanimous AI's new computer platform was put to the following test: twenty volunteers were chosen who had a knowledge of horse racing in general and the upcoming 2017 Kentucky Derby specifically. Volunteers could not have any professional or financial relationship to the horse racing industry.

At a specified time, the twenty volunteers, each at their own computers, logged onto the Unanimous AI website and were given access to a type of digital magnet controlled by a computer mouse. On the website screen was a circular list of horses entered in the upcoming Derby.

At a given signal, the volunteers moved a "graphical puck" on the screen towards the listed horse they thought would win. Each puck was influenced by a central computer server which analyzed the directional force of each user creating a "dynamic feedback loop" that influenced the overall movement of each puck. According to the founder of Unanimous AI, Louis Rosenberg,

> In this way, real-time synchronous control is enabled across a swarm (group) of distributed users. Through the collaborative control of the graphical puck, a real-time physical negotiation emerges among the networked members. This occurs because all of the participating users are able to push and pull on the puck at the same time, collectively exploring the decision-space and converging upon the most agreeable answers.[169]

The Unanimous AI computer, analyzing the collective movement of the users towards the names of the horses on the screen, posted the predicted finish, first through fourth.

The odds for correctly picking the first four finishers in exact order for that 2017 Kentucky Derby were slightly more than five hundred to one.

A journalist from TechRepublic, an online trade publication for Information Technology, was invited to watch the process at the central location where she could see the final results. Impressed with the process, she put down a $1 bet on the exact order of finish the computer identified, based on weighing the movements of all the volunteers. The team at Unanimous AI which developed the algorithm was more optimistic and put down a $20 bet.

The top four finishing horses finished in the exact order predicted.

The journalist won over $500.

The Unanimous AI team won $11,000.

It's interesting that not one of the individual participants in the test was able to correctly predict more than one horse in the right order Yet, the computer's feedback loop system that weighed the overall pattern of choices, was able to detect a group intelligence that was far greater than the intelligence of any one individual.

The following year the same test was conducted for picking Academy Award winners. Again, the volunteers all had an interest in movies, but no professional or financial relationship to the industry. The predictions using the Unanimous AI computer algorithm were 75 percent accurate compared to 65 percent accuracy by top movie critics.

In 2017, the Unanimous AI computer algorithm weighed the underlying pattern of volunteers to correctly predict *TIME* magazine's Person of the Year.

Subsequently, radiologists at Stanford University Medical Center are using Unanimous AI collaboratively to dramatically improve diagnosing pneumonia from chest x-rays. X Prize, a non-profit foundation which is a global leader in designing and implementing innovative competitions for developing new programs to benefit humanity, is using Unanimous AI to enhance collaboration.

The success of the Unanimous AI algorithm has been the subject of articles in *The Wall Street Journal*, the BBC, and *Newsweek*, among many others.

How can this Unanimous AI computer detect a greater group intelligence which none of the individual members have?

9

WELCOME *to* ONE *of* NATURE'S BRAINSTORMS: SWARM INTELLIGENCE

Look deep into nature, and then you will
understand everything better.
—ALBERT EINSTEIN

As stated on the Unanimous AI website, Swarm Intelligence is "the process that allows flocks of birds, schools of fish, and swarms of bees to reach optimal decisions with remarkable efficiency."

From our human perspective the word, "swarm" has negative connotations, eliciting visions of mob violence, not some form of deep intelligence. The power of swarm intelligence in Nature, however, takes visible form in examples such as the amazing ability of swarming bees to solve complex, life-or-death problems in an awesome way.

In order for a bee colony to survive, the selection of a location for a successful new hive is incredibly complex: the new hive has to provide the proper angle to the sun for warmth as an energy source along with proximity to water and to flowers providing nectar which the bees detect through an electric force.

A certain number of the hive community are genetically programed to be scouts. Individually these scouts fly for distances of up to thirty square miles and, upon returning to the group, engage in an oscillating "waggle dance" through which they communicate and weigh the overall assets and liabilities of potential locations. Despite the fact that no bee has any amount of discernable individual intelligence, a swarm of scout bees have a high rate of successfully choosing the right location for a sustainable hive.

Louis Rosenberg, founder of Unanimous AI, described the process in a 2021 online article:

> Biologists have shown that honeybees pick the best solution over 80 percent of the time. A human business team trying to select the ideal location for a new factory would face a similarly complex problem and find it very difficult to choose optimally, and yet simple honeybees achieve this.
>
> And unlike us humans, bees don't let arguments get in the way or succumb to whoever is loudest or most forceful. They don't split up and go off in different directions. They reach decisions that are best for the group, nature's way of combining a group's diverse perspectives with the aim of maximizing their collective wisdom.[170]

We saw how ancient bacteria, through the pathway of more novel, complex, collaborative feedback loops, developed the ability to create food sources from sunlight through photosynthesis. This pathway led to the emergence of plants and then animals using even more complex forms of swarm intelligence, such as ants using chemical traces and fish using vibrations in the water to coordinate successful group behavior. A swarm of birds can conduct accurate, sudden turns in total unison, even against powerful winds, using the feedback system generated from flapping wings.

As of this writing, no ants, fish, or birds have received a Nobel Prize for these amazing achievements done every minute of every day.

Louis Rosenberg adds, "If we consider the leap in intelligence between an individual ant and a full ant colony working as one, can we expect the same level of amplification as we go from single individual humans to an elevated '*hyper-mind*' that emerges from real-time human swarming?"[171]

In 1997, De Kerckhove, noting the emergence of the World Wide Web, predicted a related type of heightened group collaborative intelligence. As he wrote in his book, *Connected Intelligence*, "Rapid progress in intelligent software is opening avenues for the migration of psychological processes such as memory and intelligence from the inside of individual minds to the outside world of connected-knowledge media."[172]

This "migration of psychological processes ... from the inside of individual minds to the outside world of connected-knowledge media" is a theme running through our journey through **the Quantum & the Dream** synchronicity, namely, the potential gateway between the human psyche (primarily the unconscious) and the outside world, which has always existed, now potentially amplified with computer intelligence through the digital screen.

> *"Dreams never give up on us."*
> —MONTAGUE ULLMAN

I've had my own awe-inspiring personal experience of heightened group collaborations, in the first dream group I attended.

In 2003, I discovered the website of a world-renowned dream expert Dr. Montague Ullman, who had retired as emeritus clinical psychiatrist at Albert Einstein College of Medicine. He was known for an innovative form of group dream work based on his confidence that people outside the oversight of professional psychologists and psychoanalysts could safely and successfully explore the meaning of their dreams in a group setting. He went further, stating his dislike of the term "dream interpretation," preferring "dream appreciation."

There was a notice on his website that the annual dream group seminar he conducted at his home (less than two hours from me) was coming up in a few weeks. I called the phone number listed and

Ullman himself answered. After briefly introducing myself, I told him about my radio talk program and recent interest in dreamwork. He told me he restricted it to fourteen people, and it was currently filled, but would be glad to put me on a waiting list.

Three days later Monty called back to say there was now a slot open and I could have it. My excitement was off the charts.

The first day of the workshop, before Monty started, I got to meet the other participants: about 2/3 were in the field of psychology, and the remainder included a novelist and vice president of IBM. One participant had flown in all the way from Norway.

The key to Monty's method includes feeling the heart and soul of the dream rather than any specific technical interpretation or analytical method.

Here is the first dream offered by one of the participants:

"I'm lying on the ground under a tree. It's autumn. A few leaves have fallen. There's a stream nearby."

That's the entire dream.

I remember thinking to myself, "Why would anybody offer such a sparse, uninteresting dream and how the hell are we going to fill the allotted time (forty-five minutes for each dream)?"

I was in for quite a surprise. So was the dreamer.

Following Monty's method, after presenting the dream, the dreamer left the circle by moving her chair further back (we were seated on chairs in a circle).

Those of us in the circle then took turns "projecting" what the dream would mean to us if it was our dream. Monty made it clear those of us in the circle were not trying to tell the dreamer what we thought her dream meant—we were to take on the dream as if it were our own and free associate with it while the dreamer listened and, if she wanted to, took notes.

While this event was over twenty years ago, I still remember some of the projections on the dream from the circle:

"If this were my dream, I'd focus on the tree as my family tree. I'm thinking about some family issues. . . ."

"If this were my dream, I'm enjoying the relaxation of lying on the ground on a nice autumn day . . . the dream is telling me to be more relaxed about stresses going on in my life. . . ."

"If this were my dream, I have an intuitive picture of this tree as the one providing the apple Adam and Eve ate. I'm, wondering where the snake might be . . ."

I offered, "If this were my dream, I'm attracted to those few leaves which have fallen to the ground. I hear the phrase 'turning over a new leaf.'"

When the dreamer had the last word, she revealed that she had been going through one of the most stressful periods in her life, trying to decide whether or not to stay in her current job which paid very well but was uninspiring, or throw caution to the winds and pursue her love of art. She went on to tell us that a number of the projections she heard from the group led her to an awareness that her fear of risk taking came from her family upbringing (the family "tree"), where as a teenager she was encouraged not to challenge the status quo and felt guilty about expressing herself fully (Adam and Eve's guilt for eating of the "tree" of knowledge). She was now determined to "turn over a new leaf" and make the career change.

As with the "swarm intelligence" generated by Unanimous AI and the self-organizing success of thousands of volunteers at Wikipedia, Monty's dream group is a recognition of the "heightened group collaborative intelligence." While each individual clearly has a unique, important role to play, what occurs is that mystifying "greater intelligence" that emerges out of the individual projections.

Whether we engage in forms of heightened group collaborative intelligence on our own or amplify it by combining human and computer intelligence as seen with Unanimous AI, the process of shifting to the right-hemisphere capacity for imagination, empathy, and the desire for a "bigger picture" of our place in the world puts us more in synch with Nature's evolutionary drive.

In his book, *Appreciating Dreams*, Monty writes, *"Asleep and dreaming. we are in pursuit of freedom in those areas that have eluded us while awake. The relationship between freedom and truth is the driving force of our dreams."*

10

BREWING IDEAS & SPIRITED SPIRITS:
A SELF-ORGANIZED REVOLUTION

Most of what we were taught in school about the American Revolution centers around the political and military heroes—Washington, Jefferson, Adams, Franklin, and Paine—along with distinctive events such as the midnight ride of Paul Revere and the shot heard round the world at Lexington and Concord.

What's missing?

While the accomplishments of Washington, Jefferson, Adams, Franklin, Paine, etc. were decisive influences on the outcome, these individuals would never have been in a position to lead the revolution if not for the thousands of mostly spontaneous, self-organizing political, philosophical, economic, spiritual, and religious conversations, debates, and arguments, stimulated by large doses of caffeine and liquor, together with the hundreds of small printed pamphlets tacked onto walls and hawked on street corners.

Historian Bruce Richardson writes about The Green Dragon Tavern, a popular Boston venue in the years leading up to the Revolutionary War:

> It stood on Union Street, in the heart of the town's business center from 1697 to 1832, and figured in practically all the important local and national events during its long career. Red-coated British soldiers, colonial governors, bewigged crown officers, earls and dukes, citizens of high estate, plotting revolutionists of lesser degree, conspirators

in the Boston Tea Party, patriots and generals of the Revolution—all these frequently gathered at the Green Dragon to discuss their various interests over their cups of coffee, tea, and stronger drinks. In the words of Daniel Webster, this famous coffee-house tavern was the "headquarters of the Revolution."[173]

Salvatore Colleluori of George Washington University writes in his book, *The Colonial Tavern: Crucible of the American Revolution,*

> While alcohol was a prominent fixture in Colonial life, oftentimes the location where one consumed said alcohol was equally as relevant. Public houses, and more specifically taverns, played an especially important role—they weren't simply places to drink. Rather, they served as a venue to meet like-minded individuals, and functioned as clearinghouses and test beds of revolutionary ideas. . . . In taverns across the colonies, literate patriots drank and read the news of the day aloud to their fellow revelers, thereby stoking revolutionary fervor.[174]

The most powerful communications medium available for self-organizing fervor at this time were pamphlets handed out on street corners and posted in public meeting places.

Historian Bernard Bailyn writes,

> There were more than four hundred pamphlets published in the colonies on the imperial controversy up through 1776, and nearly four times that number by war's end in 1783. These pamphlets varied in their theme and approach, including tracts of constitutional theory or history, sermons

and orations, correspondence, literary pieces, and political debate. Pamphlets were one of the most important conveyors of ideas during the imperial crisis. Often written by elites under pseudonyms and published by booksellers, they have long been held by historians as the lifeblood of the American Revolution.[175]

How could any of the thousands of colonists—without any external, organized campaign—arguing, sharing information, and offering philosophical insights on freedom in those taverns, and while hawking their pamphlets on street corners, know with any certainty what grounds they were seeding and what historic fervor might be created?

They didn't know. They were participants in a self-organizing, impossible to predict, uncertain feedback loop which, as we've seen, is inherent in Nature's evolutionary pattern towards collaboration.

It was this decades-long chaotic, boisterous, collective buzz percolating through taverns, town halls, street corners, and village squares that fertilized the ground and planted the seeds for revolutionary change, out of which emerged the "Founding Fathers" and the well-known events taught in classrooms nationwide.

Postscript: The famous version of Paul Revere's heroic midnight ride was, in fact, a fictionalized tale from the imaginative mind of the poet Henry Wadsworth Longfellow. While Revere was part of a spy network which set out to warn colonial leaders in Concord that British soldiers were planning to arrest them, he never made it there, but was captured by the British and subsequently released. It was two other riders, Dawes and Prescott, who made it on horseback to Concord. Since stories are one of the most effective teaching tools, and humans are the only living organisms on the planet who invent stories, Longfellow, the right-hemisphere-oriented storyteller, wrote his poem in 1860 on the cusp of the Civil War, playing an effective role

219

in motivating patriotic fervor. Still, our educational system continues to insist the midnight ride was factual history, along with the notion that to understand this event is to focus on the military and political leaders, ignoring the all-important decades-long, self-organizing seeding of chaotic conversations and the power of imaginative story-telling as the underlying patterns for evolutionary change.

Today those chaotic tavern conversations and individually printed pamphlets have morphed into billions of emails, websites, Instagram posts, YouTube videos, etc. coursing around the planet at electric speed every second of every day.

If the New Renaissance is to fully bloom, it will be from a similar pattern of self-organizing, undirected interplay of dialogue, grievances, novel insights, and imaginative leaps, induced in large part from the intensifying pressures of escalating climate change and massively accelerating computer intelligence which are transforming the planet and humanity at speeds never before experienced.

One of the most perceptive storylines for addressing the sped-up, fast-changing, globally-connected zeitgeist is expressed in an acronym not yet well-recognized. That's where we travel next.

11

VUCA:
"HEY *it's* CRAZY OUT THERE"

In 1985, seeing the hyperspeed changes taking place globally, two prescient, "big picture" management experts, Warren Bennis and Burt Nanus, developed new strategies for increasing resilience, adaptability, and open-mindedness.

Although oriented for businesses and organizations, it was the Army War College that proactively adopted Bennis and Nanus' insights in response to the collapse of the Soviet Union and the rise of computer intelligence around the globe. Wars could no longer be fought with World War II tactics.

Today the benefits of these innovative perceptions and strategies are being adopted by organizations ranging from Harvard Business School to the Dorothy A. Johnson Center for Philanthropy. This hypermodern new vision is referred to by the acronym VUCA, which stands for:

Volatility
Uncertainty
Complexity
Ambiguity

Interestingly, we've seen all four of these qualities during our trip into the creative potential of the unconscious mind and the non-material, nonlocal, "entangled" mysteries of the quantum realm. We've seen it in the shift from the left hemisphere of the human brain to the right hemisphere, illustrated by Einstein's thought experiments chasing a light beam, Jung descending down the dark, mysterious underground staircase in his "big dream," Niels Bohr contemplating the spiritual depths of the Tao, and the numinous power to both create and destroy as symbolized by the mythic god Shiva standing on the grounds where the World Wide Web was born. (For example, when Einstein was chasing a light beam in his imaginary thought experiment, how could he have seen through the veil of the entrenched Newtonian, rational, materialist paradigm had he not been open to volatility, uncertainty, complexity and ambiguity?)

We've seen Volatility, Uncertainty, Complexity, and Ambiguity permeating the creative process of the unconscious mind, as in psychologist Julian Jaynes' description of it as a "secret theater of speechless monologue and prevenient counsel, an invisible mansion of all moods, musings, and mysteries."

We've seen volatility, uncertainty, complexity, and ambiguity in the enigmatic, mind-bending realities of quantum nature as in science historian James Gleick's quote,

> Quantum Mechanics suffused into the lay culture as a mystical fog. It was uncertainty, it was acausality, it was the Tao updated, it was the century's richest fount of paradoxes; it was the permeable membrane between the observer and the observed; it was the funny business sending shudders up science's all-too-deterministic structure.[176]

Now we see some mainstream businesses and organizations recognizing the need to think and act with more agility and non-linear creativity, adapting to the shift from both the left to the right hemispheres of the brain and from the printed page to the digital screen. As one VUCA-related website notes, echoing **the Quantum & the Dream** philosophical and depth psychological pattern, "the findings of brain research show that "the reason-seeking ratio is only a 'servant' of an intuitively—and partly unconscious—made decision."[177]

As described in an article for the *Harvard Business Review,* the VUCA strategy is a "catchall for 'Hey, it's crazy out there!'"

In the late 1980s-early 1990s—the same period Tim Berners-Lee was developing the protocols of what would become the World Wide Web—the VUCA perception of reality spread into a handful of leading business schools and organizations. To dive deeper into the patterns of VUCA strategy, here are some specific examples from a 2019 article on the subject, alternating with insights from communications theorist Marshall McLuhan, writing over 50 years ago about effective ways to recognize the deeper patterns generated by modern, electronic media:

> VUCA: "Anticipate risks but don't invest too much time in long-term strategic plans. Don't automatically rely on past solutions and instead place increased value on new, temporary solutions, in response to such an unpredictable climate."

> McLuhan: "Everything is changing—you, your family, your neighborhood, your education, your job, your government, your relation to others. And they're changing dramatically."

> VUCA: "Leverage diversity—as our networks of stakeholders increase in complexity and size, be sure to draw on the multiple points of view and experience they offer. Doing so will help you expect the unexpected."

> McLuhan: "A singular point of view can be a dangerous luxury when substituted for insight and understanding."

VUCA: "Think big picture. Make decisions based as much on intuition as analysis."

McLuhan: "Our time is a time for crossing barriers ... for looking ahead, for probing around."[178]

This last phrase, "time for crossing barriers," can be seen as both a metaphor and neuroscientific evidence of our ability to consciously shift from the primarily left-hemisphere perception which has dominated science, philosophy, psychology, and culture for centuries, across the "barrier" of the corpus collosum—the bundle of nerve fibers which anatomically separates the two hemispheres of our brains—to the larger and more intricately wired right hemisphere, much more capable of creatively and intuitively thriving in the current VUCA environment.

We've seen how the pulse of Volatility, Uncertainty, Complexity and Ambiguity permeate the realms of both the human unconscious and the quantum realm of wave-particle duality and entanglement. This is the pulse that continues to erode the materialist paradigm, a pulse we've been observing throughout our journey.

This pulse is now being recognized, slowly, but surely, in the business world. According to the consultancy firm Intelligent Management in their article, "Systems Thinking and Quantum Theory—Why We Need Them for Business" we find:

> What we are experiencing today is the full thrust of a moment of historical transition. We are transitioning from a world where phenomena could be described and managed with the knowledge and information that had been available for hundreds of years, to a new world where complexity dominates. There is a level of interconnection today that means it makes no sense to "cut things up" into pieces and try and

manage them like a machine made up of separate parts.[179]

Echoing the clarion call of right-hemisphere awareness in the age of exponentially increasing computer intelligence, the article states, **"It is vital to consider the big picture."**

(Part of this bigger picture requires an important distinction between "complexity," the "C" in VUCA, and "complication." A "complicated" situation is well served by the left-hemisphere ability to break an issue into separate parts and see how they fit together. A "complex" situation is well served by the right-hemisphere ability to contemplate a situation intuitively, all at once, like a Zen koan, allowing an answer to emerge from the creative unconscious, transcending the rational part of the brain.)

12

THE NEW RENAISSANCE:
A SHIFT INWARD

He looked at his own Soul with a telescope.
What seemed all irregular, he saw and showed
to be beautiful constellations and he added to the
consciousness hidden worlds within worlds.
—SAMUEL TAYLOR COLERIDGE

A significant symbol of the great Italian Renaissance is Galileo peering outward through his newly-improved telescope into the darkness of the solar system illuminated by celestial objects never seen before, providing a "bigger picture" of the starry heavens.

Flash forward approximately three hundred years to 1900, where Freud's publication of *The Interpretation of Dreams* and Planck's accidental discovery of the quantum, inverts Galileo's telescope, initiating the look inward, shining a light into the dark center of the human unconscious and the subatomic realm.

Galileo peering outward through his telescope confirms the shocking recognition that the sun does not revolve around the Earth, which no longer can be seen as the center of the Universe but as one of many celestial objects orbiting the sun.

Freud and, soon after, Jung, peering inward towards the deeper regions of the unconscious, reveal the ego is not the solid center of a rational, personal universe, but part of what Jung described as "a boiling cauldron of contradictory impulses, inhibitions, and affects." The early quantum theorists, peering into the inner sanctum of the quantum world, confirm the shocking recognition of contradictory impulses underlying the ordered, mechanistic, predictable Newtonian Universe.

The great Italian Renaissance artists and architects sought to portray Nature's sense of harmony and proportion.

The great cubist painters, such as Picasso, and influential surrealist painters, such as Dali, sought to capture the enigmatic depths of the unconscious dream and the mystifying paradox of quantum wave-particle duality and entanglement.

The humanist philosophers of the Italian Renaissance celebrated the dignity and power of individual thought and curiosity after one thousand-plus years of church-imposed bans on free thinking and forced capitulation to its authority.

In the twenty-first century, influenced by the emergence of the World Wide Web, a number of forward-thinking philosophers, social critics, and scientists, while advocating for individual, contemplative thought, at the same time recognize the need for a post-humanist consciousness, one generated by heightened group collaborative intelligence. Now that we are connected through a planetary digital Web, what happens globally now affects our individual nervous systems.

13

CROSSING *the* STARGATE

Three of the most effective ways we humans learn are:

- Trial and Error
- Games (play)
- Stories

The first, trial and error, is a key for all living organisms (interesting that our educational system actually penalizes us for making errors, thereby inhibiting the learning process).

As for the second way we learn, until relatively recently, it was thought only mammals engaged in play as a learning activity. As confirmed in a *Discover* magazine article, "It's not just kittens and baby chimps that play, but also birds, reptiles, fish and even invertebrates, including spiders and wasps. We have reports of octopuses fooling around with Lego blocks and Komodo dragons waging tug of war with their keepers. In 2015, a study of tooth marks on fossils showed that the bones may have served as a toy for a tyrannosaurid more than sixty-five million years ago."[180]

It's the third activity, stories, that distinguish human learning from all others on this planet. As we've seen along the trip through **the Quantum & the Dream** synchronicity, certain stories transcend time and place, rising to the level of wisdom tales and enduring myths.

FINAL THOUGHT EXPERIMENT: PIVOTAL SCENE IN *2001: A SPACE ODYSSEY*:

With the issue of powerful supercomputers permeating human life here in the twenty-first century, a storyline well worth looking at is the memorable scene in *2001: A Space Odyssey* when, after HAL refuses to accept a technical mistake it made and learns the two astronauts on board plan to unplug it, it kills one and attempts to kill the other. Bowman, having survived HAL's attempt, gets inside its brain and shuts it down, circuit by circuit.

It's easy to understand why most viewers see this scene as part of the long tradition of heroic journeys where the protagonist has to win a physical battle with a powerful, deadly enemy. It's easy to see the dismantling of HAL's computer brain as a warning of the impending threat of supercomputers in real life here on Earth.

Considering Dorothy's arrival as a totally new heroic archetype within **the Quantum & the Dream** synchronicity, one based on empathy and collaboration, more in tune with what we've seen as Nature's inherent drive towards novel, more complex collaboration, an alternative meaning can be seen in Bowman's need to shut down HAL's brain.

HAL's advanced supercomputer brain, as with today's supercomputers, is already far superior to our brains at left-hemisphere attributes such as researching, organizing, and analyzing information. It was also far superior at overseeing the space ship's complex mechanical systems, which is why the two astronauts are shown to have relatively little to do in getting the spaceship to Jupiter's orbit. Without HAL, the first part of the mission, getting to the source of the monolith's signal, could not have been achieved.

The second part of the mission, however—to make contact with this higher intelligence by crossing the mysterious, transcendent stargate in preparation for a complete physical, emotional, and spiritual transformation into the star child—could not be assisted by HAL's left-hemisphere intelligence. Once crossing through the boundary-exploding, hyper-dimensional, beyond the known fabric of space-time Stargate, Bowman's brain was stretched to a shocking level (as evidenced by his contorted face in the movie). To resist a complete cognitive meltdown required the total shift from left hemisphere analysis and search for certainty to the boundary-expansive, imaginative, "big picture" capacity of the right hemisphere.

In this interpretive version, the scene showing Bowman floating through the interior of HAL's brain, disengaging the circuitry, can be seen as preparation to switch from the left-hemisphere skillset—technically getting the space ship to the precipice of the monolith's signal—to the right-hemisphere open-minded descent into the unknown required for any significant inner transformation to occur.

In Bowman's case, his ability to flow with rather than resist the unrecognizable process orchestrated by the higher intelligence led to his rebirth as a star child.

A rebirth is the literal definition of a renaissance.

If, in this thought experiment, the monolith (higher intelligence) is seen as a symbol for Nature's Intelligence and its inherent drive to create novel, more complex, and collaborative organisms, then the sequence of HAL bringing Bowman to the precipice of the stargate, to Bowman's dismantling HAL's brain, to Bowman's crossing the stargate to be transformed—becomes a storyline connected to the proposed formula where HI (human Intelligence) has shifted its consciousness towards the right hemisphere of the brain.

$$\text{NI}\left(\frac{\text{HI}}{+}\atop{\text{AI}}\right) = \text{New Renaissance}$$

©

Postscript 1: As previously noted, when Max Planck unintentionally discovered the mysterious, discontinuous quantum—which flew in the face of science's accepted Newtonian view of mechanical motion and Maxwell's well-defined description of light, revealing the paradox of wave-particle duality—his left-hemisphere need for pragmatic, unambiguous clarity made him so uncomfortable, he resisted making his discovery public. To his great credit, after two decades of collaborative conversations with Einstein, Bohr, Heisenberg, Pauli, et al., Planck shifted to a right-hemisphere, transcendent vision worthy of Bowman's experience through the Stargate, as revealed in his later statement: "Science enhances the moral value of life, because it furthers a love of truth and reverence . . . reverence, because every advance in knowledge brings us face to face with the mystery of our own being."

What a powerful, spiritually-infused phrase: "Face-to-face with the mystery of our own being."

Planck's transformative vision, centering on "the moral value of life," "reverence," and coming "face to face with the mystery of our own being" are wonderful prescriptions for ushering in the New Renaissance.

Postscript 2: Bowman's hyper-speed, unpredictable, disorienting experience in crossing the stargate has many of the volatile, uncertain, complex, and ambiguous sensations we are all now, to various extents, feeling in the current zeitgeist. Bowman's experience reflects Marshall

McLuhan's descriptions of the effects of modern media extending our nervous systems outwards, anticipating the global connectedness of the World Wide Web. As McLuhan said, "Ours is the Age of Implosion, of inclusive consciousness and deep personal involvement ... we enter the age of the unconscious ... faced with information overload, we have no alternative but pattern-recognition."

It is this act of pattern recognition which has been at the heart of our trip through **the Quantum & the Dream** synchronicity, a pattern emerging from the depths of the collective unconscious.

The very title of Kubrick's movie reveals its connection to Homer's ancient Greek masterpiece *The Odyssey*. The connection is enhanced by naming the heroic astronaut "Bowman" (Odysseus was known for his skill as an archer or bowman). Just as astronaut Bowman had to cross the mysterious, mind-staggering stargate to complete his mission, so Odysseus had to descend into Hades, the dark, mysterious Underworld, to learn from the blind prophet Tiresias the true purpose of his mission.

We can see ourselves "entangled" in this same archetypal drama of transformation—just as Odysseus was often blown off course by the strong winds of Zeus, so we, here in the twenty-first century, are being blown off course by the strong winds of global warming, wreaking havoc across the planet.

For the New Renaissance we need to add one more crucial element to the crossing of the threshold. Homer's *The Odyssey* and Kubrick and Clarke's *2001: A Space Odyssey* are male-driven storylines. We need to recognize Dorothy's arrival in the same year as Freud's book on dreams and Planck's discovery of the quantum, a new female-hero archetype, percolating up from the collective unconscious, whose strength comes from empathy, loyalty, and heart as she crosses the threshold into the uncertain, transcendent world of the dream.

Carl Jung showed us how the powerful storylines of the Hero's Journey are embedded in the depths of the collective unconscious which "contains the whole spiritual heritage of mankind's evolution, born anew in the brain structure of every individual."

For it is out of the collective unconscious that Bowman, Dorothy, and Odysseus emerge, as well as Shiva, poised in its cosmic dance on the grounds of CERN, to help us detect the deep pattern of change at the core of Nature's Intelligence.

RIFFING *on the* NEW RENAISSANCE

There are many dramatic visions of what the future will look like. These run the gamut from Google Director of Engineering, Ray Kurzweil's prediction that within thirty years the human consciousness will literally be uploaded into computer servers, overcoming the limitation of aging, to Nobel Prize-winning physicist Stephen Hawking's dystopic, "The development of full artificial intelligence could spell the end of the human race."

Perhaps Homo sapiens will use genetic engineering and quantum computing to morph into a totally different species, as different from us as we are from apes.

I share the optimism of cultural historian William Irwin Thompson. Back in the 1990s, when I was asked who is the one person I haven't yet interviewed but would like to, my answer was William Irwin Thompson. My wish was granted a little over twenty years ago, about a year before my dream group experience with Montague Ullman.

Bill, who died in 2020, was a philosopher, cultural historian, and founder of the Lindisfarne Association, a network of scientists, artists, and religious scholars who collaborated on what Bill called "Planetary Culture." Listening to Thompson or reading his work, one gets the sense he has not only read virtually every important book in the humanities and general sciences, but has integrated them into a unique, wholistic vision. He was truly a renaissance thinker.

What I remember most about our two-hour radio conversation wasn't the content of what he said, but his style, which he himself described as "poetic, philosophical mind-jazz."

Here are some of Bill's insights from his essay, "It's Already Begun, The Planetary Age":

> We live in a culture that we do not see. We don't live in industrial civilization; we live in planetization. For example, we all think we live in a world that's structured according to industrial nation-states that engage in activities of trade and warfare that are weighed and measured by certain quantitative forms. That's the conscious structure of the world that we call reality. The unconscious structure of the world is that there are all kinds of forms of dark exchange called pollution – atmospheric things like acid rain and the greenhouse effect and changes in the oceans – and that these are the integrations that are bringing us all together. We are in an implosive situation of planetary integration, but where is the planetary culture expressed?[181]

This last statement sounds contradictory. How could such massive, potentially deadly issues such as acid rain and toxic oceans be "integrations?" Thompson was a close colleague of Lynn Margulis and a strong advocate of the Gaia theory she cocreated with James Lovelock. Margulis viewed the growing pollution caused by humans to be a natural part of the Gaian system since we are the progeny of evolution's drive. She and Thompson, both "big picture" thinkers, see the expanding awareness of the environmental destruction we are causing as a signal being emitted calling on homo sapiens to finally "wise up," transcend the materialistic paradigm, and create a "planetary consciousness." What makes this situation complex is that, as Thompson pointed out, "We live in a culture we do not see" and "the unconscious structure of the world" puts us in an "implosive situation of planetary integration."

(It's interesting that Thompson should use the same term, "implosive," as we saw recently in McLuhan's insight: "Ours is the Age of Implosion, of inclusive consciousness and deep personal involvement . . . we enter the age of the unconscious . . . faced with information overload, we have no alternative but pattern-recognition.")

The *"implosion"* they refer to can be seen as starting in the year 1900 synchronicity when Freud's book on dreams, Baum's story of Dorothy's descent into the dream, and Planck's discovery of the quantum realm "imploded" the rational, mechanistic, all-too-certain materialistic paradigm.

Thompson expands on the vision of a planetary culture, also confirming McLuhan's insights on the age of sped-up, electronic media:

> It's expressed with those people who are sensitive to
> the unconscious, who live at the membrane between
> the culture's conscious system, called civilization and
> writing and literacy, and those who face the intuitive
> dimensions of the unconscious. These are the artists.
> These are the prophets. . . . There is also planetary
> culture in the forms of electronic communication, of
> the whole grid of satellites that enables us to exist in
> forms that have nothing to do with the reality of the
> industrial nation-state.[182]

Following Thompson's cultural insights, the prediction I would propose regarding the New Renaissance is a new vision of what it means to be educated. The cemented brick walls of my old elementary school, among all the others, will continue to erode as people figure out that the most effective education in the era of the World Wide Web is a classroom without walls, no longer controlled by central school authorities stuck in dusty, antiquated mindsets based on controlling students, not opening up their minds. People will start self-organizing, meeting in local, supervised groups, where students interact in person and connect through digital screens with other local groups of supervised students, all tapping into open-source knowledge/wisdom gateways on the Web (a process already expanding on websites such as Coursera and Skillshare).

One of the most significant insights for seeding the New Renaissance is to contemplate the underlying meaning of "educare" and "sapiens."

Educare, the original Latin word on which "education" is based, as previously observed, means **"to draw out."**

Sapiens, the word our ancestors chose for classifying us humans, means **"wisdom."**

The current age, with the escalating, potentially apocalyptic devastation of global warming and hyper-fast computers increasing their intelligence and influence every day, now calls upon whatever capacity we can develop to "draw out wisdom." This must happen both individually— through thought experiments, dreams, deep contemplation, flashes of insight, and imaginative leaps, connecting to the deeper wisdom inherent in our unconscious mind—and collectively, in what William Irwin Thompson referred to as "noetic polities:" communities organized around inner transformation, spiritual insight, and the goal of encouraging a more collaborative civilization, a project much more achievable in the age of the World Wide Web than when Thompson was working at it back in the 1980s. He writes,

> The hope for such rapid and thoroughgoing cultural transformations is now more possible than ever because we presently live in a "noetic polity" based on the continuous exchange of ideas and instantaneous flow of information crisscrossing virtually the entire planet. We must finally bring the freedom of our

imagination to bear on what the shape of things to come may yet turn out to be because only the imagination is really big enough and wild enough to entertain the unthinkable possibilities beyond the ideas and information that currently rule and define our world.[183]

Integrating this new form of drawing out wisdom with a sense of play will be most helpful in creating this New Renaissance. As media theorist Derrick De Kerckhove writes, "It has often been observed that playing and learning are born genetically intertwined in us, only to be artificially unraveled at school."[184]

In the current age, hyper-speed changes have totally disrupted the mindset which Baby Boomers grew up with: get through school, work at the same job or career for forty or so years. No more. Today, according to the Bureau of Labor Statistics, "the average number of jobs in a lifetime is twelve."[185]

McLuhan, who first opened our eyes to the inseparability of psyche and media, anticipating the global, intertwined network of the Web and the eventual crumbling of the cement walls of antiquated education, predicted—using his favorite style, the playful pun—that in the future people wouldn't be earning a living, but "learning a living."

Where does it say we should stop learning at any time we're alive? Out of the evolutionary drive of Nature's Intelligence, we were given the capacity to keep increasing our capacity to learn. Maybe that's the main reason we're here in the first place.

CODA:
TOWARDS *a* BIGGER PICTURE

From Galileo peering outward through his telescope, to Freud, Baum, and Planck inverting it to look inward, to recognizing Gaia weaving her web into novel, more complex collaborations, it's been quite a ride.

Some postcards from our trip:

- Einstein chasing an imaginary light beam;
- Dorothy descending into the dreamscape of Oz;
- Jung, deep into his big dream, lifting a stone slab and descending that mysterious staircase;
- Niels Bohr, integrating his vision of the mystical Tao with the inner world of the quantum;
- Dali's surreal pocket watches melting in the sun;
- Lynn Margulis descending into the world of bacteria to discover Evolution's inherent drive towards novel, more complex collaboration;
- Astronauts on a moon mission capturing a photographic image of the entire Earth (Gaia), a transcendent image which becomes one of the most reproduced images in history;
- Tim Berners-Lee, determined to enhance connected intelligence, developing the protocols for what becomes the World Wide Web at the CERN, the world's largest particle accelerator;
- The statue Shiva arriving at CERN, displaying the cosmic dance of creation and destruction where the World Wide

Web was born and where the huge particle accelerator continues its search into the mysteries of the quantum realm.

Does it feel as if Evolution, Nature's Intelligence, is moving us closer to one of its punctuated leaps?

If a handful of fourteenth-century Italian philosophers, artists, and patrons could stare into the face of the Black Plague, on its way to wiping out one third of all Europeans, rich and poor alike, and induce a great, transformative renaissance at a time when the only way to mass-produce knowledge and wisdom was through the laborious hand-copying of manuscripts for the small minority of literate citizens, then why can't we, in the twenty-first century, stare into the ravages of escalating climate change and bring forth the New Renaissance at a time when accessing the greatest knowledge and wisdom, both ancient and contemporary, are just a few keyboard taps away through AI-driven digital screens accessed by billions around the planet?

If a thin, fragile blade of grass can push through a tight pore in a cement sidewalk through its inherent drive to seek sunlight, can't we push through our fears and anxieties to bring forth the light of a New Renaissance?

GETTING IN SYNC

At the heart of the universe is a steady, insistent beat: the sound of cycles in sync. It pervades nature at every scale from the nucleus to the cosmos.

—STEVEN STROGATZ, AUTHOR, SYNC: HOW ORDER EMERGES FROM CHAOS IN THE UNIVERSE, NATURE, AND DAILY LIFE

At the heart of this journey is a synchronicity, a term created by Carl Jung referring to two or more events which clearly didn't cause each other, but are filled with too much meaning to be just a coincidence. He poetically described synchronicity as "a falling together in time."

Let's be on the lookout for individuals taking on the heroic archetypes of Odysseus, Dorothy, and Bowman, willing to descend into the depths of their being, returning to help weave the next great storyline. They're out there. To find them requires getting through the dissonant noise of the twenty-four-hour news cycle increasing every day.

One of the most memorable statements made on my radio program comes from the dynamic Buddhist teacher Robert Thurman: "Enlightenment is tolerance of cognitive dissonance."

Beneath the dissonance is the "steady, insistent beat" of Nature's Intelligence continuing her drive towards novel, expansive collaboration.

A final insight from another brilliant woman, anthropologist Margaret Meade, whose career brought so much knowledge and wisdom about us humans: "Never doubt that a small group of thoughtful, committed citizens can change the world; indeed, it's the only thing that ever has."

ENDNOTES

SHIFT ONE

1   Iain McGilchrist, *The Master and his Emissary: The Divided Brain and the Making of the Western World* (New Haven: Yale University Press, 2010), 22.

2   Ned Herrmann, "Is it true that creativity resides in the right hemisphere of the brain?", *Scientific American.com* (1998).

3   McGilchrist, *The Master and his Emissary*, 187.

4   John S. Rigden, "Einstein's Revolutionary Paper," *Physics World.com* (April 1, 2005).

5   Walter Isaacson, "The Light-Beam Rider," *New York Times Section SR* (October 20, 2015): 6.

6   Dale M. Kushner, "Dreams and Our Need for Empathy and Imagination," *Dale M. Kushner* (blog), (December 1, 2016).

7   As quoted by Dan Falk, "A Debate over the Physics of Time," *Quantum Magazine* (July 19, 2016).

8   McGilchrist, *The Master and his Emissary*, 40.

9   Steven Aftergood, "Schopenhauer and Unconscious Thought," *FAS. org*, (February 22, 2006).

10   Rosie Lesso, "The Impact of Sigmund Freud's Theories on Art," *The Collector.com* (May 15, 2020).

11   Stephen B. Parker, "1909: Jung Descends into the Collective Unconscious," *Jung Currents.com* (2018).

12   Parker, "1909: Jung Descends."

13   Carl Jung, *The Structure and Dynamics of the Psyche* (Princeton: Princeton University Press, 1970), 342.

14 McGilchrist, *The Master and his Emissary*, 40.

15 Thomas J. McFarlane, "Quantum Physics, Depth Psychology, and Beyond," *integralscience.org* (February 26, 2000).

16 Carl Jung, *Letters of C. G. Jung, Volume 2, 1951-1961* (Abingdon: Routledge, 1976), 108-109.

17 Carl Jung, Wolfgang Pauli, *The Interpretation of Nature and the Psyche* (New York: Pantheon Books, 1955), 29.

18 Fritjof Capra, "Heisenberg and Tagore," *Fritjof Capra.net* (July 3, 2017).

19 Carlo Rovelli, *Helgoland: Making Sense of the Quantum Revolution* (New York: Riverhead Books, 2021), 156.

20 Montague Ullman, "The Dream: In Search of a New Abode," Presented at the twenty-third Annual Meeting of the International Association for the Study of Dreams, July 22, 2006.

21 David Z. Albert and Riva Galchen, "Was Einstein Wrong? A Quantum Threat to Special Relativity," *Scientific American.com* (March 1, 2009).

22 Brian Greene, *The Fabric of the Cosmos: Space, Time, and the Texture of Reality* (New York: Vintage, 2005).

23 Greene, *The Fabric*.

24 Betony Adams, "Do quantum effects play a role in consciousness?" *Physics World.com* (January 26, 2021).

25 Materials provided by National Institute of Standards and Technology, "Team builds quantum hybrid system by entangling molecule with atom," *Science Daily.com* (May 20, 2020).

26 Erwin Schrödinger, *AZquotes.com* (March 16, 2013).

27 Princeton University Press, "William R. Newman on Newton the Alchemist," *Princeton University Press* (November 7, 2018).

28 Michael Woronko, "Carl Jung, Alchemy, and Quantum Physics, with Murray Stein," *A Philosopher's Stone* (April 9, 2020).

29 Rovelli, *Helgoland*, 75-76.

30 McGilchrist, *The Master and his Emissary*, 187.

31 Julian Jaynes Society, "Julian Jaynes's The Origin of Consciousness in the Breakdown of the Bicameral Mind," *Julian Jaynes Society* (2023).

32 F. David Peat, "Wolfgang Pauli: Resurrection of Spirit in the World," *F. David Peat.com* (accessed March 3, 2023).

33 Peat, "Wolfgang Pauli."

34 Maria Popova, "Atom, Archetype, and the Invention of Synchronicity: How Iconic Psychiatrist Carl Jung and Nobel-Winning Physicist Wolfgang Pauli Bridged Mind and Matter," *The Marginalian* (March 9, 2017).

35 Amanda Gefter, "Why Two Geniuses Delved into the Occult," *New Scientist* (April 24, 2009).

36 The Dali Museum, "Freud," *Dali.org* (accessed on April 14, 2023).

37 Carmen Ruiz, "Salvador Dali and Science: Beyond a Mere Curiosity," *Centre for Dalinian Studies* (2010).

38 Carl Jung, *Synchronicity: An Acausal Connecting Principle* (Princeton: Princeton University Press, 2010), 29.

39 PBS, "Heisenberg States the Uncertainty Principle 1927," *PBS.org* (1998).

40 Dali Universe, "A Brief History of the Surrealist Image—The Persistence of Memory," *thedaliuniverse.com*.

41 Emma Taggart, "5 Masters of Surrealism Who Painted Their Dreams and Visualized Their Inner Minds," *My Modern Met.com* (July 3, 2021).

42 F. David Peat, "Synchronicity," *F. David Peat.com*.

43 McGilchrist, *The Master and his Emissary*, 3.

44 C. G. Jung, *Aion: Researches into the Phenomenology of the Self* (Abingdon: Routledge, 1991), 9ii, paragraph 126.

45 Arthur Zajonc, *Catching the Light: The Entwined History of Light and Mind* (Oxford: Oxford University Press, 1993), 12.

46 McFarlane, "Quantum Physics."

47 Tom Snyder TV interview of Marshall McLuhan, *The Tomorrow Show NBC*, 1976.

48 Fritjof Capra Interview, *fritjovcapra.net* (1985).

49 Fritjof Capra, quoted at *awaken.org* (accessed on April 18, 2023).

50 SuperSummary, "Dancing Wu Li Masters," *supersummary.com* (accessed on March 3, 2023).

51   Krishnamurti Foundation Trust, "Bohm & Krishnamurti," *kfoundation.org* (accessed on April 18, 2023).

52   The Krishnamurti Foundation, "Teachings," *jkrishnamurti.org* (August 7, 1964).

53   CERN, "Where did it all begin?" *home.cern* (accessed on March 3, 2023).

54   Frank Close, Martin Michael, and Christine Sutton, *The Particle Odyssey: Journey to the Heart of the Matter* (Oxford: Oxford University Press, 2002), 228.

55   Hern, Alex, "Tim Berners-Lee on 30 years of the world wide web: 'We can get the web we want,'" *theguardian.com* (March 12, 2019).

56   Academy of Achievement, "Tim Berners-Lee: Father of the World Wide Web," *achievement.org* (October 18, 2018).

57   Better Help Editorial Team, "Exploring Free Association," *Betterhelp.com* (October 6, 2023).

58   Tim Berners-Lee as quoted in "Tim Berners-Lee's dream for the web," *semanticabyss.blogspot.com* (blog) May 25, 2009.

59   Fritjof Capra, "Shiva's Cosmic Dance at CERN," *fritjofcapra.net* (June 20, 2004).

60   Aiden Randle-Conde, quoted in "Maha Shivratri: Here's why the world's largest particle physics lab CERN has Shiva's 'Nataraj' statue," *dnaindia.com* (March 4, 2019).

61   Wolf-Dieter Storl, "Shiva: The Wild God of Power and Ecstasy," *intuitive-connections.net* (accessed on March 8, 2023).

SHIFT TWO

62   Plato, as quoted in Marshall McLuhan, *The Medium is the Message* (New York: Bantam Books, 1967) 113.

63   Vannevar Bush, "As We May Think," *theatlantic.com* (July, 1945).

64   Fritjof Capra, *The Web of Life: A New Scientific Understanding of Living Systems* (New York: Anchor Books, 1996), 35.

65   Eric McLuhan and Frank Zingrone, eds., *Essential McLuhan* (Abingdon: Routledge 1997), 273.

66 M. McLuhan, *The Medium is the Message*, 4.

67 As quoted in "Q&A with Stephen Hawking," *usatoday.com* (December 2, 2014).

68 Ferris Jabr, "The Reading Brain in the Digital Age: The Science of Paper Versus Screens," *scientificamerican.com* (April 11, 2013).

69 Jabr, "The Reading Brain."

70 Kevin Kelly, T*he Inevitable: Understanding 12 Technological Forces That Will Shape Our Future* (New York: Penguin Books, 2017), 88.

71 Bush, "As We May Think."

72 Katherine Hayles, "Hyper and Deep Attention: The Generational Divide in Cognitive Modes," *researchgate.net* (accessed October 19, 2023).

73 Marshall McLuhan, *Counterblast* (Toronto: McClelland and Stuart Limited, 1969), 41.

74 Rose Joel, "How to Break Free of Our 19th Century Factory-Model Education System," *Atlantic Magazine* (May 9, 2012).

75 Melissa Gouty, "Why You Need to Develop a Biliterate Brain—and How to Do It," *literaturelust.com* (August 19, 2020).

76 Gouty, "Why You Need."

77 Linda Poppenheimer, "Paper versus Digital Media—Environmental Impact" *greengroundswell.com* (April 10, 2017).

78 19 Eco Friendly Habits, "Paper Wastage Facts: Statistics about Paper Waste You Must Know," *ecofriendlyhabits.com* (2021).

79 Shaikat Hossain, "The Internet as a Tool for Studying the Collective Unconscious," *Jung Journal: Culture and Psyche* (Spring 2012): 103-109.

80 Ephrat Livni, "The Internet is a manifestation of our psyches—neither better nor worse," *qz.com* (November 8, 2018).

81 Werner Geyser, "16 Real Social Media Stats You Should Pay Attention to," *influencermarketinghub.com* (August 2, 2023).

82 Chad Guzman, "The Facebook Whistle Blower Revealed Herself on *60 Minutes*. Here's What You Need to Know," *Time.com* (October 4, 2021).

83 Carl Jung, as quoted in "Jung Lexicon," *jungpage.org* (October 27, 2013).

84 Roxanne Gay, "Why People Are So Awful Online," *nytimes.com* (July 17, 2021).

85 Elaine McArdie, "Oh, What a Tangled Web We Weave," *hls.harvard. edu* (July 7, 2021).

86 Yuval Noah Harari, quoted at *bestquotes.top* (accessed on October 20, 2023).

87 Elizabeth Hartney, "The Symptoms and Risks of Television Addiction," *verywellmind.com* (February 10, 2022).

88 L. R. Huesmann et al, "Early Exposure to TV Violence Predicts Aggression in Adulthood," *apa.org* (2003).

89 Vaughan Bell, "Don't Touch That Dial," *slate.com* (February 15, 2010).

90 Farhad Manjoo, "The Moral Panic Engulfing Instagram," *nytimes.com* (October 13, 2021).

91 Thomas Moore, *Care of the Soul: A Guide for Cultivating Depth and Sacredness in Everyday Life* (New York: Harper Perennial, 1994).

92 Tim Berners-Lee, "Where Does the World Wide Web Go From Here?" *wired.com* (March 11, 2019).

93 Marshall McLuhan, *Understanding Media: The Extensions of Man* (New York, New American Library, 1964), 3.

94 M. McLuhan, *Understanding Media*, 3-4.

95 Tim Parks, "The Chattering Mind," *nybooks.com* (June 29, 2012).

96 Edward Bernays, quoted at *historytoday.com* (February 6, 2019).

97 McFarlane, "Quantum Physics."

ORIENTATION

98 Jeremy Taylor, *Dream Work: Techniques for Discovering the Creative Power in Dreams* (Ramsey, NJ: Paulist Press, 1983), 7-8.

99 DailyHistory.org, "How Did the Bubonic Plague Make the Italian Renaissance Possible?" *dailyhistory.org* (September 22, 2021).

100 Antonio Guterres as quoted at *press.un.org* (November 1, 2021).

101 NASA Global Climate Change at *climate.nasa.gov* (accessed on October 20, 2023).

102 Damian Carrington, "Air pollution is slashing billions of years off the lives of billions, report finds," *theguardian.com* (September 1, 2021).

103 Jordan Davidson, "Air Pollution Responsible for over 6.6 Million Deaths Worldwide, Study Finds," *ecowatch.com* (October 21, 2020).

104 Kenneth V. Rosenberg et al., "Decline of the North American Avifauna," *science.org* (September 19, 2019).

105 Marco Margaritoff, "The American Bumblebee Is About to Become an Endangered Species for the First Time in History," *allthatsinteresting. com* (October 12, 2021).

106 Sites at Penn State, "Overpopulation" (blog), *sites.psu.edu* (January 12, 2018).

107 The World Counts, "Hectares of forests cut down or burned," *theworldcounts.com* (accessed on October 20, 2023).

108 Sites at Penn State, "Overpopulation" (blog).

109 Graham Peebles, "Overpopulation, food waste, and climate change," *nationofchange.org* (February 8, 2021).

110 Sites at Penn State, "Overpopulation" (blog).

111 Sites at Penn State, "Overpopulation" (blog).

112 Angela Dewan, "Germany's deadly floods were up to 9 times more likely due to climate change, study estimates," *cnn.com* (August 24, 2021).

113 New York University, "Flooding significantly impacts food security, new study finds," *phys.org* (October 17, 2022).

114 Livia Albeck-Ripka, Thomas Fuller, and Jack Healy, "The Ashes of the Dixie Fire Cast a Pall 1,000 Miles from Its Flames," *nytimes.com* (October 11, 2021).

115 Jessica Colarossi, "How Global Warming Helped Ignite One of Australia's Worst Fire Seasons," *bu.edu* (January 14, 2020).

116 Victoria Milko and Julie Watson, "UN: Climate change to uproot millions, especially in Asia," *apnews.com* (March 2, 2022).

117 Dr. Cheryl Mathews, "Anxiety Disorder Statistics," anxietyhub.org (August 22, 2017).

118 "Anxiety Disorders—Facts & Statistics," Anxiety & Depression Association of America (October 28, 2022).

119 Matt Richtel, "Emergency Room Visits Have Risen Sharply for Young People in Mental Distress, Study Finds," *nytimes.com* (May 1, 2023).

120 Diane Archer, "Nearly half of older households have no retirement savings," *justcareusa.org* (April 2, 2019).

121 Michelle Cottle, "Getting Old is a Crisis More and More Americans Can't Afford," *nytimes.com* (August 9, 2021).

SHIFT THREE

122 Madelaine Bohme, Rudiger Braun, and Florian Breier, "Nature's Masterpiece: How Evolution Gave Us Our Human Hands," *discovermagazine.com* (November 13, 2020).

123 Robin Kundis Craig, "Learning to Live with the Trickster: Narrating Climate Change and the Value of Resilience Thinking," *law.arizona. edu* (Spring 2016).

124 Clea Simon, "Can tech save us from worst climate change effects? Doesn't look good," *news.harvard.edu* (November 17, 2022).

125 Simon, "Can tech save us."

126 Adrienne LaFrance, "The Coming Humanist Renaissance," *The Atlantic* (July/August 2023).

127 LaFrance, "The Coming Humanist Renaissance."

128 Elizabeth Letts as quoted in "Elizabeth Letts Finds the Wizard Behind Oz," *publishersweekly.com* (January 4, 2019).

129 Fritjof Capra, *The Tao of Physics: An Explanation of the Parallels between Modern Physics and Eastern Mysticism* (Boulder, CO: Shambhala, 1975), 325.

130 L. Frank Baum, excerpted at *penguinrandomhouse.ca* (accessed on October 23, 2023).

131 John Steinbeck as quoted at *quotessqyings.net* (accessed on October 21, 2023).

132 Adam Hadhazy, "Why Is Our Universe Filled with Spirals?" *discovermagazine.com* (May 17, 2019).

133 Carl Jung, *Psychology and Alchemy* (Princeton: Princeton University Press, 1980), 325.

134 European Molecular Biology Laboratory, "Spiral Growth: Feedback Loop Behind Spiral Patterns in Plants Uncovered?" *sciencedaily.com* (Nov 3, 2016).

135 *Tao te Ching*, Verse 78, as quoted at *harinam.com* (accessed on October 21, 2023).

136 Derrick De Kerckhove, *Connected Intelligence: The Arrival of the Web Society* (South Brisbane: Somerville House Publishing, 1997), 148.

137 Herbert Spencer as quoted at *simple.wikipedia*.org,(accessed on October 21, 2023).

138 Lynn Margulis and Dorion Sagan, *Dazzle Gradually: Reflections On the Nature of Nature* (White River Junction, VT: Chelsea Green Publishing, 2007), p 37.

139 Margulis and Sagan, *Dazzle Gradually*, 40.

140 Margulis and Sagan, *Dazzle Gradually*, 37.

141 Margulis and Sagan, *Dazzle Gradually*, 45.

142 Margulis and Sagan, *Dazzle Gradually*, 45.

143 Capra, *The Tao of Physics*, 68.

144 Margulis and Sagan, *Dazzle Gradually*, 96.

145 Margulis and Sagan, *Dazzle Gradually*, 5.

146 Margulis and Sagan, *Dazzle Gradually*, 5.

147 Elisabet Sahtouris, "After Darwin," *radical.org* (August 30, 2003).

148 Sahtouris, "After Darwin."

149 Sahtouris, "After Darwin."

150 Sahtouris, "After Darwin."

151 Richard Grant, *"Do Trees Talk to Each Other,"* Smithsonianmag.com *(March, 2018).*

152 Grant, "Do Trees Talk."

153 Elisabet Sahtouris and Willis W. Harman, *Biology Revisioned* (Berkeley: North Atlantic Books, 1998), 15.

154 Sahtouris and Harman, *Biology Revisioned*, 36.

155 Sahtouris and Harman, *Biology Revisioned*, 198.

156 Elisabet Sahtouris, "Prologue to a New Model of a Living Universe," *sahtouris.com* (accessed on October 22, 2023).

157 Denise Hruby, "To Stop an Extinction, He's Flying High, Followed by his Beloved Birds," *nytimes.com* (August 8, 2023).

158 Fritjof Capra, as quoted at *libquotes.com* (accessed on October 22, 2023).

159 Yuval Noah Harari as quoted in "How Humankind Can Become Totally Useless," *time.com* (February 16, 2017).

160 Dario Taraborelli, "Seven years after Nature, pilot study compares Wikipedia favorably to other encyclopedias in three languages," *diff.wikimedia.org* (August 2, 2012).

161 Michael Blanding, "Wikipedia or Encyclopedia Britannica: Which has More Bias?" *forbes.com* (January 20, 2015).

162 Kelly, *The Inevitable*, 272.

163 Kelly, *The Inevitable*, 273.

164 De Kerckhove, *Connected Intelligence*, 152.

165 Jon Gertner, "Wikipedia's Moment of Truth," *nytimes.com* (July 18, 2003).

166 De Kerckhove, *Connected Intelligence*, 160.

167 La France, "The Coming Humanist Renaissance."

168 William Irwin Thompson, as quoted at *theyalsosaidso.com* (accessed on October 22, 2023).

169 Louis Rosenberg, "Human Swarming and the Future of Collective Intelligence," *singularityweblog.com* (July 20, 2015).

170 Louis Rosenberg, "Swarm Intelligence: AI inspired by honeybees can help us make better decisions," *bigthink.com* (October 13, 2021).

171 Rosenberg, "Human Swarming."

172 De Kerckhove, *Connected Intelligence*, 143.

173 Bruce Richardson, "Coffee Houses in Colonial Boston," *bostonteapartyship.com* (accessed on October 22, 2023).

174 Salvatore Colleluori, "The Colonial Tavern, Crucible of the American Revolution," *warontherocks.com* (April 17, 2015).

175 Robert G. Parkinson, "Print, the Press, and the American Revolution," *oxfordre.com* (September 3, 2015).

176 James Gleick, quoted at *epdf.pub* (accessed on October 22, 2023).

177 Waltraud Glaser, "Where does the term "VUCA" come from?" (blog) *vuca-world.org* (accessed on April 23, 2023).

178 Joyoti Kukreja, "Holacracy: The Next Generation of Leadership in a VUCA World," *ijtsrd.com* (October, 2019).

179 Intelligent Management, "Systems Thinking and Quantum Theory— Why We Need Them For Business" (blog) *intelligentmanagement.ws* (April 8, 2021).

180 Marta Zarasca, "Non-mammals Like to Play Too," *discovermagazine. com* (November 21, 2019).

181 William Irwin Thompson, "It's Already Begun: The Planetary Age is an unacknowledged daily reality," *context.org* (accessed on October 23, 2023).

182 Thompson, "It's Already Begun."

183 Ralph Peters, "The Gaian Politics of Lindisfarne's William Irwin Thompson," *earthlight.org* (accessed October 23, 2023).

184 De Kerckhove, *Connected Intelligence*, 17.

185 Bureau of Labor Statistics, "Number of Jobs, Labor Experience, Marital Status, and Health for Those Born 1957-1964," *bls.gov* (August 22, 2023).

ACKNOWLEDGMENTS

To Peter Rogen: Your enthusiasm, support, and guidance for the ideas and perception of this book were paramount.

To Cait Johnson, a great editor and spiritual presence.

To Gail Torr, a passionate publicist who believed in this book.

To Ed Rosenfeld and Neal Goldsmith for your encouragement and providing a salon space where I could test out many of the ideas and insights for this book.

To Jerry and Sasha Gillman for starting such a creative radio station and including me on the team, and to Gary Chetkof for keeping it going.

To Victoria Sullivan and Ron VanWarmer, my radio co-hosts, for helping to keep the conversations going.

To humanities professor Leo Brody and philosophy professor Charles Frankel for opening up my mind to the creative depths of the human imagination.

To dream experts Montague Ullman and Jeremy Taylor for all of those transcendent dream group experiences.

To Susan Rosen for all those group conversations at Miriam's Well.

To my parents and grandparents, who, despite whatever darkness appeared, knew how to laugh out loud.

INDEX